Easy
Small Gardens

Easy
Small Gardens

Gardening successfully in small spaces

PETER McHOY

PREMIER

First published in 1999 by
Hermes House

© Anness Publishing Limited 1999

Hermes House is an imprint of
Anness Publishing Limited
Hermes House
88-89 Blackfriars Road
London SE1 8HA

ISBN 1 84038 373 9

A CIP catalogue record for this book is available from the British Library

Publisher: Joanna Lorenz
Senior Editor: Caroline Davison
Designer: Ian Sandom
Jacket Designer: Michael Morey
Production Controller: Mark Fennell
Commissioned Photography: John Freeman
Original Design: Patrick McLeavey & Partners

Previously published as part of a larger compendium, *The Small Garden Book*

Printed and bound in Italy

1 3 5 7 9 10 8 6 4 2

PAGE 1: A wall-mounted fountain is perfect if you
do not have room for a whole pond.
PAGE 2: A collection of containers can brighten
up a small courtyard.
PAGE 3: A well-planted container makes an eye-
catching focal point.
PAGE 4: *Hydrangea anomala petiolaris*.
PAGE 5: This wall provides a textured backdrop
for an abundance of brightly coloured flowers.

CONTENTS

INTRODUCTION

SMALL GARDENS CAN STILL HAVE A BIG IMPACT, and this book shows you how to get the best from a limited area. Size is comparative, of course, and if your gardening is confined to a backyard or terrace, or just a tiny balcony, even a small town garden can seem large. On the other hand, a medium-sized country garden may appear small to the owner of a large estate. But whether you garden on a rooftop, have a fairly typical small town garden, or a larger plot that simply seems small for you – and all that you would like to achieve within its confines – this book offers plenty of inspirational ideas and practical solutions to help you design your perfect small garden.

Making the most of an outdoor area depends partly on good design and partly on suitable planting. If you want a low-maintenance garden the emphasis should be on the hard landscaping and the use of ground cover and low-maintenance plants. If you are a plant collector, a design with the emphasis on the planting space will be important … but choosing the right plants in proportion to the available space is vital.

You will find lots of ideas for redesigning a garden, from initial ideas to execution. But sometimes only minor modifications to your existing garden are necessary for a complete transformation, and the section on features and structures has plenty of thought-provoking ideas that you might like to consider.

In the third section of the book, there is helpful advice on designing beds and borders such as island beds and one-sided borders. There is also guidance on creating a neat edge to a border, on marking out oval and curved beds, and on planting up the border. Further advice on planting shrubs, herbaceous plants and ground cover plants should provide you with the confidence and know-how to design your own stunning beds and borders.

OPPOSITE: *Pots and containers are a good means of incorporating herbs in a very confined space.*

Size is always an issue when designing a garden, so it is important to choose the right plants for the size of your garden, and to remember that plants grow at different rates. Herbaceous plants, for example, will take only a few years to reach their mature height, but slower-growing trees and shrubs will take many more years. You should also bear in mind that the dimensions given in books can be no more than a crude guide. Heights can vary greatly, according to soil, position and local climate. Some trees and shrubs call be kept compact by regular pruning – for example, buddleias and eucalyptus both grow quite tall if left unpruned, but will make compact shrubs with a good shape if cut back severely each spring.

Although you will find plenty of suggestions in the following pages, attractive gardens are not designed to a rigid formula, and there is always room for individual interpretation – and even eccentricity. Some gardens are designed to shock, some are traditional in concept, a few are strictly formal, and many are a compromise between formality and informality. There are as many styles as there are tastes, and the only criterion for success is whether the result pleases you personally.

Design does not become easier with decreasing size: rather, it becomes more difficult and demanding. A large garden tends to look good anyway, with the odd weedy bed going almost unnoticed among the overall impression of large lawns, stately trees and shrubs. In a small garden long vistas are out of the question and the use of trees and large shrubs is often severely limited. Every part of the garden comes under the spotlight, and errors of judgement are often emphasized.

All these handicaps call be overcome, however, and, as the illustrations in this book show, you can transform a currently uninspiring plot into something quite magical and unique.

ELEMENTS *of* DESIGN

*Attractive small gardens seldom just happen, they are designed.
And despite the apparent contradiction, the smaller the garden,
the more important good design becomes. A small garden can be taken in
almost at a glance, and the difference between good and bad design,
attention to detail or neglect, is immediately obvious. In a small garden
there is always the temptation to try to cram in many more features
than there is really space for. Keep the design simple, stick to a style,
and follow the suggestions in this section for making a plan to scale.
Then check the effect by marking out the shapes in the garden
before you start work. This way you will be assured of success.*

ABOVE: *Sometimes accommodating essentials
attractively, such as this tool shed, can become a
problem. Careful screening can help minimize
the effect they have on the garden.*

OPPOSITE: *A small garden should not lack
impact. Provided it is well planted and has
some strong focal points, it becomes easy to
ignore the limitations of size.*

PLANNING YOUR GARDEN

SOME SUCCESSFUL GARDENS ARE WORKED OUT on the ground, in the mind's eye, perhaps visualized during a walk around the garden, or conceived in stages as construction takes place. This approach is for the gifted or very experienced, and it is far better to make your mistakes on paper first.

A major redesign can be time-consuming and expensive, especially if it involves hard landscaping (paving, walls, steps, etc). However, simply moving a few plants is rarely enough to transform an uninspiring garden into something special. It is worth having a goal, a plan to work to, even if you have to compromise along the way. Bear in mind that you may be able to stagger the work and cost over several seasons,

but having a well thought out design ensures the garden evolves in a structured way.

Use the checklist opposite to clarify your 'needs', then decide in your own mind the *style* of garden you want. Make a note of mundane and practical considerations, like where to dry the clothes and put the refuse, plus objects that need to be screened, such as a compost area, or an unpleasant view.

Unattractive views, and necessary but unsightly objects within the garden, such as toolsheds, are a particular problem because they can dominate a small garden. Well-positioned shrubs and small trees can act as a screen. To improve the outlook instantly use a large plant in a tub.

ABOVE: *In this garden the bird table helps to draw the eye away from the practical corner of the garden.*

LEFT: *Make a small garden look larger than it really is by ensuring the sides are well planted and creating a striking focal point.*

OPPOSITE: *Shape and form can be as important as colour in creating a stylish garden.*

LABOUR-SAVING TIPS

● To minimize cost and labour, retain as many paths and areas of paving as possible, but only if they don't compromise the design.

● If you want to enlarge an area of paving, or improve its appearance, it may be possible to pave over the top and thus avoid the arduous task of removing the original.

● Modifying the shape of your lawn is easier than digging it up and relaying a new one. It is simple to trim it to a smaller shape if you want a lawn of the same area, and if you wish to change the angle or shape, it may be possible to leave most of it intact, and simply lift and relay some of the turf.

LEFT: *Strong lines and several changes of level give this small garden plenty of interest. In this kind of design, the hard landscaping is more important than the soft landscaping (the plants).*

CHOICES CHECKLIST

Before you draw up your design, make a list of requirements for your ideal garden. You will almost certainly have to abandon or defer some of them, but at least you will realize which features are most important to you.

Use this checklist at the rough plan stage, when decisions have to be made . . . and it is easy to change your mind!

Features

Barbecue	☐
Beds	☐
Borders, for herbaceous	☐
Borders, for shrubs	☐
Borders, mixed	☐
Birdbath	☐
Changes of level	☐
Fruit garden	☐
Gravelled area	☐
Greenhouse/conservatory	☐
Herb garden	☐
Lawn (mainly for decoration)	☐
Lawn (mainly for recreation)	☐
Ornaments	☐
Patio/terrace	☐
Pergola	☐
Pond	☐
Raised beds	☐
Summerhouse	☐
Sundial	☐
Vegetable plot	☐
Plus	☐

Functional features

Compost area	☐
Garage	☐
Toolshed	☐
Plus	☐

Necessities

Children's play area	☐
Climbing frame	☐
Sandpit	☐
Swing	☐
Clothes dryer	☐
Dustbin area	☐
Plus	☐

CHOOSING A STYLE

Before sitting down with pencil and paper to sketch out your garden, spend a little time thinking about the style that you want to achieve. In many gardens plants and features are used for no other reason than that they appeal; an excellent reason, perhaps, but not the way to create an overall design that will make your garden stand out from others in the street.

The styles shown in the following six pages are not exhaustive, and probably none will be exactly right for your own garden, but they will help you to clarify your thoughts. You should know roughly what you want from your garden before you start to design it.

FORMAL APPROACH
Formal gardens appeal to those who delight in crisp, neat edges, straight lines and a sense of order. Many traditional suburban gardens are formal in outline, with rectangular lawns flanked by straight flower borders, and perhaps rectangular or circular flower beds cut into them. Such rigid designs are often dictated by the drive for the car and straight paths laid by the house builder.

Although the gardens shown here are all very different, what they have in common is a structure as important as the plants contained within it. The designs are largely symmetrical, with no pretence at creating a natural-looking environment for the plants.

The very size and shape of most small gardens limits the opportunities for natural-looking landscapes, so a formal style is a popular choice.

Parterres and knot gardens
Parterres and knot gardens often appeal to those with a sense of garden history, though in a small garden the effect can only ever be a shadow of the grand designs used by sixteenth-century French and Italian gardeners.

Parterres are areas consisting of a series of shaped beds, or compart-

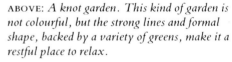

ABOVE: *A knot garden. This kind of garden is not colourful, but the strong lines and formal shape, backed by a variety of greens, make it a restful place to relax.*

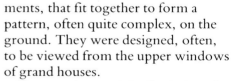

LEFT: *This small, enclosed courtyard garden balances a central focal point with a boundary that features this dramatic entrance.*

ments, that fit together to form a pattern, often quite complex, on the ground. They were designed, often, to be viewed from the upper windows of grand houses.

Knot gardens, originally designed to be viewed from above, are similar but low-growing clipped hedges are used to form the geometric and often interwoven designs. The space between hedges can be filled with flowers or, more historically correct, coloured sands or gravel, or even crushed coal if black appeals.

These are expensive gardens to create, slow to establish, and labour-intensive to maintain, but the results can be stunning. This kind of garden is unsuitable for a young family.

Formal herb gardens

Herb gardens are popular features and are much easier to create than knot gardens. Illustrations of both old and new herb gardens in books will often give you ideas for designs.

Rose gardens

A formal rose garden is easy to create, and it will look good even in its first season. To provide interest throughout the year, edge the beds with seasonal flowers and underplant the roses with spring bulbs or low-growing summer flowers.

Paved gardens

A small garden lends itself to being paved throughout. By growing most plants in raised beds or in containers, less bending is involved and many of the smaller plants are more easily appreciated. Climbers can be used to make the most of vertical space, and if you plant in open areas left in the paving, the garden can still look green.

Courtyard gardens

Space can be at a real premium in the heart of a town, but you can turn your backyard into an oasis-like courtyard garden, with floor tiles and white walls that reflect the light. Add some lush green foliage, an 'architectural' tree or large shrub, and the sound of running water. Although the plants may be few, the impact is strong.

Traditional designs

A small formal garden, with a rectangular lawn, straight herbaceous border, and rose and flower beds is still a popular choice with gardeners looking for the opportunity to grow a wide variety of plants such as summer bedding, herbaceous plants, and popular favourites such as roses. The design element is less important than the plants.

LEFT: *The use of white masonry paint can help to lighten a dark basement garden or one enclosed by high walls.*

BELOW: *This long, narrow plot has been broken up by strong lines: a useful design technique.*

INFORMAL EFFECTS

The informality of the cottage garden and the 'wilderness' atmosphere of a wild garden are difficult to achieve in a small space, especially in a town. However, with fences well clothed with plants so that modern buildings do not intrude, an informal garden can work even here.

Cottage gardens

The cottage garden style is created partly by design and the use of suitable paving materials (bricks for paths instead of modern paving slabs), and also by the choice of plants.

Relatively little hard landscaping is necessary for a cottage garden – brick paths and perhaps stepping-stones through the beds may be enough. It is the juxtaposition of 'old-fashioned' plants and vegetables that creates the casual but colourful look associated with this type of garden.

Mix annuals with perennials – especially those that will self-seed such as calendulas and *Limnanthes douglasii*, which will grow everywhere and create a colourful chaos. If flowers self-sow at the edge of the path, or between other plants, leave most of them to grow where they have chosen to put down roots.

Plant some vegetables among the flowers, and perhaps grow decorative runner beans up canes at the back of the border.

Wildlife gardens

A small wildlife garden seems almost a contradiction in terms, but even a tiny plot can offer a refuge for all kinds of creatures if you design and plant with wildlife in mind.

Wildlife enthusiasts sometimes let their gardens 'go wild'. However, this is not necessary. A garden like this one looks well kept and pretty, yet it provides long vegetation where animals and insects can hide and find

RIGHT: *The house itself will inevitably dominate a small garden, especially when you look back towards it. Covering the walls with climbers will help it to blend in unobtrusively.*

food. There is water to attract aquatic life, and flowers and shrubs to bring the butterflies and seeds for the birds.

An orchard can also be a magnet for wildlife of many kinds.

Woodland gardens

A woodland effect is clearly impractical for a very tiny garden, but if you have a long, narrow back garden, trees and shrubs can be used very effectively. Choose quick-growing deciduous trees with a light canopy (birch trees, *Betula* species, are a good choice where there's space,

RIGHT: *The woodland effect can be delightfully refreshing on a warm spring or summer day, but works best with trees that have a tall canopy that allows plenty of light to filter through. Although a pond is attractive in this situation, care will have to be taken to remove leaves in the autumn.*
BELOW: *A pretty pond is a super way to attract wildlife, and looks especially good if well integrated into the garden like this one.*

but they can grow tall). Avoid evergreens, otherwise you will lose the benefit of the spring flowers and ferns that are so much a feature of the traditional woodland garden.

Use small-growing rhododendrons and azaleas to provide colour beneath the tree canopy, and fill in with ground cover plants, naturalized bulbs such as wood anemones and bluebells, and plant woodland plants such as ferns and primroses.

Use the woodland effect to block out an unattractive view or overlooking houses. As an added bonus it is low-maintenance too.

Rocks and streams

Rock or water features alone seldom work as a 'design'. They are usually most effective planned as part of a larger scheme. Combined, however, rocks and water can be used as the central theme of a design that attempts to create a natural style in an informal garden.

Meandering meadows

Instead of the rectangular lawn usually associated with small gardens, try broadening the borders with gentle sweeps, meandering to merge with an unobstructed boundary if there is an attractive view beyond. If the distant view is unappealing, take the border round so that the lawn curves to extend beyond the point of view. Use shrubs and lower-growing border plants to create the kind of border that you might find at the edge of a strip of woodland.

Bright beds and borders

If plants are more important than the elements of design, use plenty of sweeping beds and borders, and concentrate heavily on shrubs and herbaceous plants to give the garden shape. Allow plants to tumble over edges and let them grow informally among paving.

If you want to create a strong sense of design within such a plant-oriented small garden, use focal points such as ornaments, garden seats or birdbaths.

BASIC PATTERNS

Having decided on the *style* of garden that you want, and the *features* that you need to incorporate, it is time to tackle the much more difficult task of applying them to your own garden. The chances are that your garden will be the wrong size or shape, or the situation or outlook is inappropriate to the style of garden that you have admired. The way round this impasse is to keep in mind a style without attempting to recreate it closely.

If you can't visualize the whole of your back or front garden as, say, a stone or Japanese garden, it may be possible to include the feature as an element within a more general design.

STARTING POINTS

If you analyse successful garden designs, most fall into one of the three basic patterns described below, though clever planting and variations on the themes almost always result in individual designs.

Circular theme

Circular themes are very effective at disguising the predictable shape of a rectangular garden. Circular lawns, circular patios, and circular beds are all options, and you only need to overlap and interlock a few circles to create a stylish garden. Plants fill the gaps between the curved areas and the straight edges.

Using a compass, try various combinations of circles to see whether you can create an attractive pattern. Be prepared to vary the radii and to overlap the circles if necessary.

Diagonal theme

This device creates a sense of space by taking the eye along and across the garden. Start by drawing grid lines at 45 degrees to the house or main fence. Then draw in the design, using the grid as a guide.

Rectangular theme

Most people design using a rectangular theme – even though they may not make a conscious effort to do so. The device is effective if you want to create a formal look, or wish to divide a long, narrow garden up into smaller sections.

Circular theme

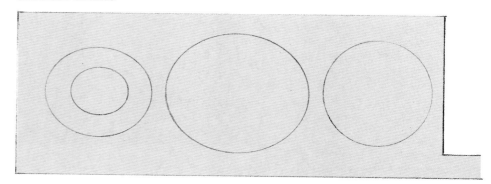

Diagonal theme

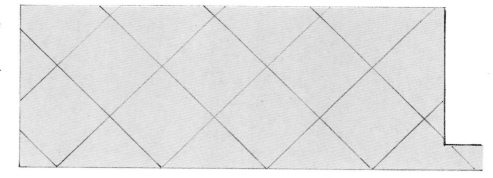

Rectangular theme

Circular theme

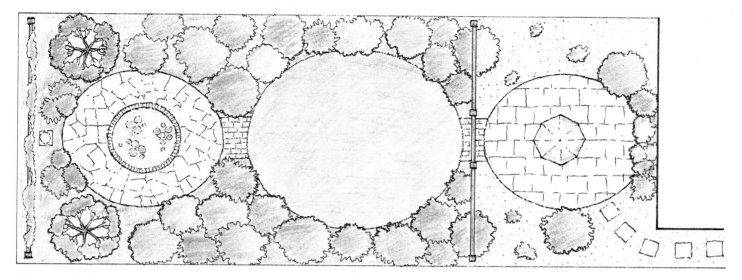

Diagonal theme

Rectangular theme

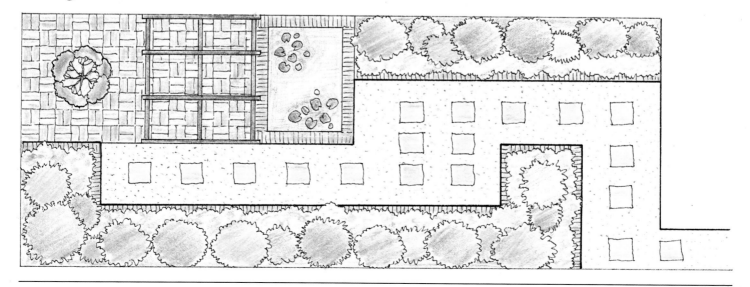

MEASURING UP

Whether designing a garden from scratch or simply modifying what you already have, you need to draw a plan of the garden as it is. A drawn plan will enable you to see the overall design clearly, and to experiment with different ideas before committing yourself to a definite option.

HOW TO MEASURE THE SITE

YOU WILL NEED:

- One, or ideally two, 30m (100ft) tape measures (unless your garden is very short). Plasticized fabric is the best material as linen stretches and steel is difficult to manipulate.
- A steel rule about 1.8m (6ft) long (to measure short distances).
- Pegs to mark out positions, and meat skewers to hold one end of the tape in position if working alone.
- Clip–board and pad or graph paper.
- A couple of pencils, sharpener and an eraser.

1 Make a rough visual sketch by eye. It does not have to be accurate, but try to keep existing important features roughly in proportion. Leave plenty of space on the plan for adding dimensions. If necessary, use several sheets of paper, and indicate where they join.

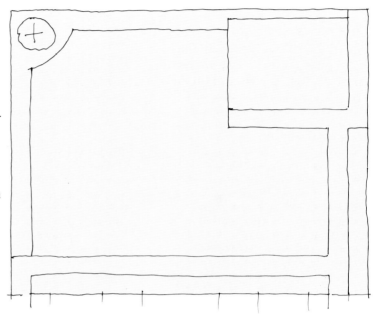

2 Choose a base line from which to start measuring. Make it a long, straight edge from which the majority of other points can be measured. A long fence or a house wall are often convenient starting points. From the straight edge or base line, measure off key points, such as the positions of windows, doors, any outbuildings, and so on. Measure out at right angles to establish the distances from the base line to the important features so that you can build the outline plan. Most key points on your plan can be established by measuring again at right angles from these right angles if necessary.

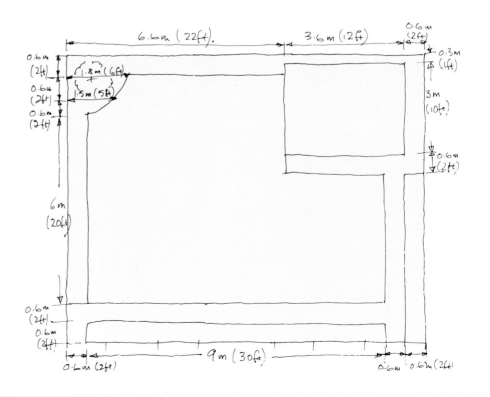

HOW TO MAKE A SCALE DRAWING

1 To make a scale drawing, choose a scale that enables you to fit the garden (or at least a self-contained section of it) onto the one sheet of graph paper. Buy large sheets of graph paper if necessary. For most small gardens, a scale of 1:50 (2cm to a metre or ¼in to 1ft) is about right. If your garden is large, try a scale of 1:100. Draw your base line in first, then transfer the scale measurements. When the right-angle measurements have been transferred, draw in the relevant outlines.

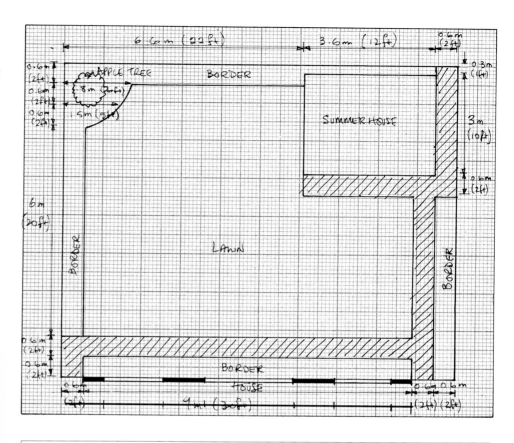

DON'T MAKE WORK

When measuring your plot, don't waste time measuring and plotting the position of features that you have no intention of retaining in your replanned garden. If you intend to remove an unsightly tree or large shrub, or to pull down a garden shed that has seen better days, leave them off your plan – they will only clutter and confuse.

SLOPES AND CONTOURS

In a large garden, slopes are often significant and may have to be taken into account. You can generally ignore gentle slopes in a small garden, or make a mental note of them.

HOW TO USE TRIANGULATION

It may not be possible to position some features or key points simply by measuring a series of right angles. These are best determined by a process known as triangulation. Using a known base, perhaps the corners of the house, simply measure the distance from two points to the position to be established. By transferring the scale distances from the two known points later, the exact position can be established. To transfer the triangulated measurements, set a compass to each of the scale distances in turn, and scribe an arc in the approximate position. Where the second arc intersects the first one, your point is established.

To fix position of tree, measure to A, then to B. Strike arcs on a scale drawing with compasses set at these measurements. Where the arcs cross shows the position of the tree in relation to the house.

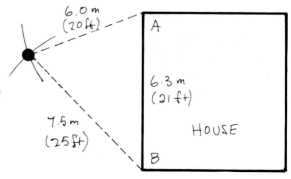

CREATING THE DESIGN

The most off-putting part of drawing up the design is the blank sheet of paper. Once this is overcome, producing alternative plans becomes fun, and there is the satisfaction of marking out the space to see the final effect in the garden. Just follow the stages below.

Stage 1: the basic grid
With the measurements transferred to graph paper you should already have a plan of your garden, showing any permanent structures and features that you want to retain.

Now superimpose onto this grid the type of design you have in mind – one based on circles, rectangles or diagonals, for example. If you are sure of the type of layout you want, draw these directly onto your plan in a second colour. If you think you might change your mind, draw the grid on a transparent overlay. For most small gardens, grid lines 1.8–2.4m (6–8ft) apart are about right.

Using an overlay, or a photocopy of your plan complete with grid, mark on the new features that you would like to include, in their approximate positions. You might find it helpful to cut out pieces of paper to an appropriate size and shape so that you can move them around.

Stage 2: the rough
Using an overlay or a photocopy, start sketching in your plan. If you can visualize an overall design, sketch this in first, then move around your features to fit into it. If you have not reached this stage, start by sketching in the features you have provisionally positioned – but be prepared to adjust them as the design evolves.

You will need to make many attempts. Don't be satisfied with the first one – it may be the best, but you won't know this unless you explore other options.

Don't worry about planting details at this stage, except perhaps for a few important plants that form focal points in the design.

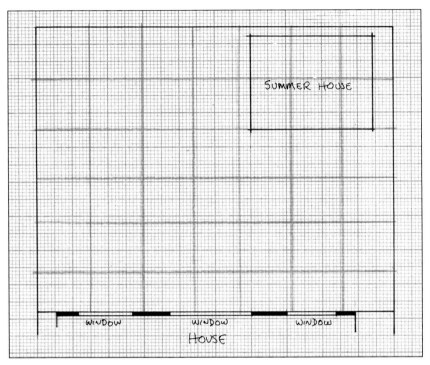

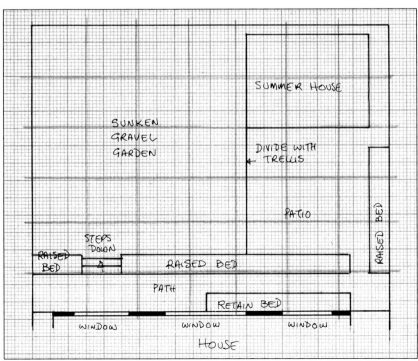

Stage 3: the detailed drawing

Details such as the type of paving should be decided now – not only because it will help you to see the final effect, but also because you need to work to areas that use multiples of full blocks, slabs or bricks if possible. Draw in key plants, especially large trees and shrubs, but omit detailed planting plans at this stage.

Trying it out

Before ordering materials or starting construction, mark out as much of the design as possible in the garden. Use string and pegs to indicate the areas, then walk around them. If possible take a look from an upstairs window. This will give a much better idea of the overall design and whether paths and sitting areas are large enough.

Use tall canes to indicate the positions of important plants and new trees. This will show how much screening they are likely to offer, and whether they may become a problem in time. By observing the shadow cast at various parts of the day, you'll also know whether shade will be a problem – for other plants or for a sitting-out area.

CONSTRUCTION

You can employ a contractor to construct the garden for you, but many gardeners prefer to get help with the main structural features, such as patios and raised beds, and do the rest of the work themselves to keep the cost down. Even the 'heavy' jobs are well within the ability of most gardeners with modest DIY skills. For more information see the next section.

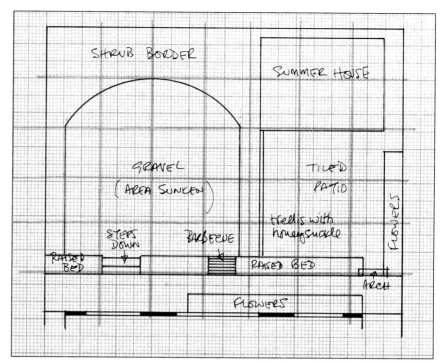

DIFFICULT SITES

DIFFICULT SITES AND PROBLEM SHAPES CAN BE A challenge, but one that can be met with a little determination and a touch of inspiration. Some ways to tackle a selection of special areas are suggested in the following pages.

If your garden is little more than a roof or a balcony, or your house has been wedged in on a building plot that is perhaps L-shaped, or even triangular, traditional garden design techniques might seem difficult to apply.

Many of the design ideas outlined in the previous chapter can still be applied, however, although you may require an alternative design strategy for specific areas.

Patios usually feature as an element in a larger overall design, but in turn have to be designed themselves. Difficult sites like slopes, windy

ABOVE: *When your front garden is as tiny as this, compensate by making the most of vertical space with climbers and windowboxes.*

LEFT: *High walls, which would otherwise have dominated this garden, are balanced by strong vertical lines. Even the tops of the walls have been put to good use!*

alleys and passageways between houses demand thoughtful planning and appropriate plants.

Front gardens present a special problem, not because of size or shape, but because a large portion of the garden is usually dedicated to the car – often there is a broad drive to the garage or a hard standing area where the vehicle is left for long periods. Legal restrictions about what you can do with your front garden can be another potential problem – especially on estates where the developers or local authority want to maintain an 'open plan' style.

If conditions really are too inhospitable for permanent plants, or the space too limited for a 'proper' garden, containers can provide the answer. Use them creatively, and be prepared to replant or rotate frequently so that they always look good, whatever the time of year.

Unpromising backyards and basements can be transformed as much by a coat of masonry paint, a few choice plants, and some elegant garden furniture and tubs, as by an extensive – and expensive – redesign. Imagination and inspiration are the keynotes for this type of garden design.

In this chapter you will find many solutions to specific problems like these, and even if your particular difficulty is not covered exactly, you should be able to find useful ideas to adapt.

ABOVE: *This long, narrow plot has been broken up into sections, with an angled path so that you don't walk along the garden in a straight line.*

LEFT: *Roof gardens are always cramped, but by keeping most of the pots around the edge it is possible to create a sense of space in the centre.*

UNUSUAL SHAPES

Turn a problem shape to your advantage by using its unusual outline to create a garden that stands out from others in your street. What was once a difficult area to fill will soon become the object of other gardeners' envy because of its originality.

Long and narrow – based on a circular theme

This plan shows a design based on a circular theme. The paved area near the house can be used as a patio, and the one at the far end for drying the washing, largely out of sight from the house. Alternatively, if the end of the garden receives more sun, change the roles of the patios.

Taking the connecting path across the garden at an angle, and using small trees or large shrubs to prevent the eye going straight along the sides, creates the impression of a garden to be explored.

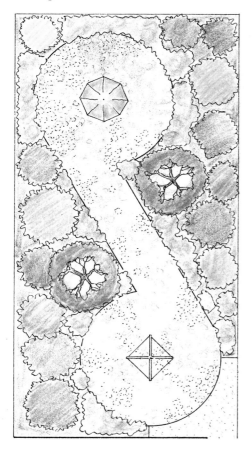

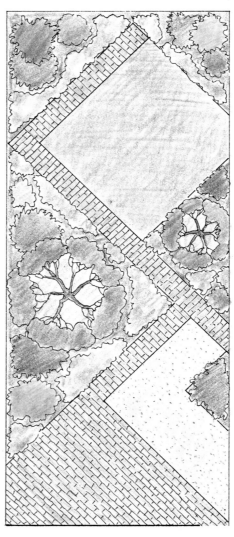

Long and narrow – based on diagonal lines

This garden uses diagonals to divide the garden into sections, but the objective is the same as the circular design. It avoids a straight path from one end of the garden to the other, and brings beds towards the centre to produce a series of mini-gardens.

Long and tapered to a point

If the garden is long as well as pointed, consider screening off the main area, leaving a gateway or arch to create the impression of more beyond while not revealing the actual shape. In this plan the narrowing area has been used as an orchard, but it could be a vegetable garden.

Staggering the three paved areas, with small changes of level too, adds interest and prevents the garden looking too long and boring. At the same time, a long view has been retained to give the impression of size.

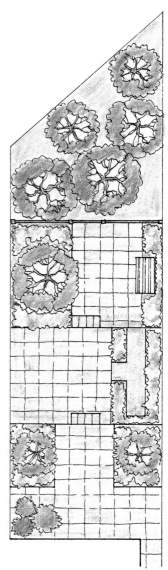

Corner sites

Corner sites are often larger than other plots in the road, and offer scope for some interesting designs. This one has been planned to make the most of the extra space at the side of the house, which has become the main feature rather than the more usual back or front areas.

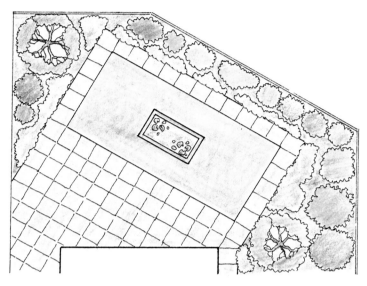

Square and squat

A small square site like this offers little scope for elaborate design, so keep to a few simple elements. To give the impression of greater space the viewpoint has been angled diagonally across the garden. For additional interest, the timber decking is slightly raised creating a change of level. In a tiny garden a small lawn can be difficult to cut, but you could try an alternative to grass, such as chamomile, which only needs mowing infrequently.

A variety of styles have been used in this plan, a combination of diagonals and circles – both of which counter the basic rectangle of the garden itself.

Curved corner sites

Curved corner gardens are more difficult to design effectively. In this plan the house is surrounded by a patio on the left-hand side, and a low wall partitions the patio from the rest of the garden, making it more private. For additional interest, the drive is separated from the gravel garden by a path. Gravel and boulders, punctuated by striking plants such as phormiums and yuccas, effectively marry the straight edges with the bold curve created by the corner site.

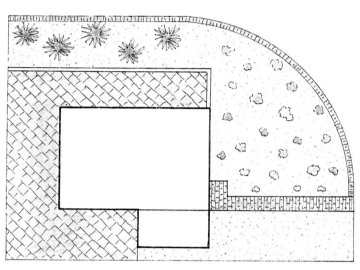

L-shaped

L-shaped gardens offer plenty of scope. Even in a small garden, the opportunity to walk around and explore an area that cannot be seen from one place is a considerable plus-point. This plan shows the clever use of focal points – a tree seat and a seat at the far end – to create a reason to explore the garden. The patio area is partially covered with overhead beams and separated from the rest of the garden by raised flowerbeds.

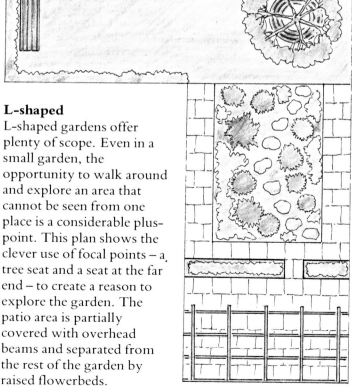

PLANNING PATIOS

The majority of small garden patios are little more than a paved area adjoining the back of the house, usually with little sense of design and often boring for most of the year. Your patio can be a key focal point that looks good in all seasons. A patio needs careful designing. It should be an attractive feature in its own right yet still form an integrated part of the total garden design.

Siting a patio

The natural choice for a sitting-out area is close to the house, especially if you plan a lot of outdoor eating. It's convenient, and forms an extra 'room', a kind of extension to the home, with a good view of the rest of the garden.

However, this spot may be shady for much of the day, in a wind tunnel created by adjacent buildings, or simply not fit in with your overall garden design.

Be prepared to move the patio away from the main building to gain

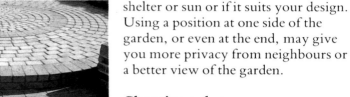

ABOVE: *Consider alternatives to paving slabs – bricks, clay and concrete pavers.*
BELOW: *The clever patio overhead makes this area function like an extra room.*

shelter or sun or if it suits your design. Using a position at one side of the garden, or even at the end, may give you more privacy from neighbours or a better view of the garden.

Choosing a shape

Most patios are rectangular – the logical shape for most gardens – but feel free to express yourself in a way that suits the overall design. A circular or semi-circular patio can form part of a circular theme. However, a round patio in a small garden designed around rectangles is likely to look incongruous.

Setting the patio at an angle to the house retains the convenience of straight lines, yet creates a strong sense of design. Consider using this shape on a corner of the house.

Patio boundaries

A clearly defined boundary will emphasize the lines of a design based on a rectangular grid. A low wall, designed with a planting cavity, will soften the hard line between paving and lawn.

High walls should be used with caution as a patio boundary, but occasionally they can be useful on one or perhaps two sides of the patio as a windbreak or privacy screen. A screen block wall will break up the space less than a solid wall, blocks or bricks. Planting suitable shrubs in front of the wall will soften the impact and help to filter the wind.

Changes of level

If the garden slopes towards the house a change of level helps to make a feature of a patio. Use a few shallow

LEFT: Patios don't have to be by the house. A cosy corner of the garden can even be more appealing.

BELOW: A patio at its best where plants and people meet. The use of bricks instead of large slabs gives the illusion of size.

steps to act like a 'doorway' to the rest of the garden.

A raised patio is a practical solution if your garden slopes away from the house. This creates a vantage point, a terrace from which you can overlook the rest of the garden. On a flat site, simply raising the level by perhaps 15cm (6in) can be enough to give the patio another dimension.

Paving materials

The choice of paving sets the tone of the patio: brash and colourful, muted but tasteful, integrated or otherwise. Do not be afraid to mix materials. Single rows of bricks will break up a large area of slabs. Choose any combination of materials that is appropriate for the setting.

If the patio is close to the house, choose bricks or pavers that match the house bricks closely. The facing bricks used for the house may be unsuitable for paving, but you should be able to achieve a close match.

FRONT GARDENS

Front gardens greet visitors and can give delight to passers-by. Unfortunately they are difficult to design well if you have to accommodate a driveway for the car, and possibly a separate path to the front door. Even enthusiastic gardeners with delightful back gardens are often let down by an uninspired front garden. We have taken four typical front gardens and shown how they can be improved. Pick ideas from any of these that you think could enhance your own space.

EXAMPLE ONE

This is a typical design for a front garden: a rectangular lawn is edged with a flower border used mainly for seasonal bedding, and bordered by a hedge. The redesigned garden concentrates on softening the harsh demarcation between drive and ornamental section. Plants now play a more prominent role, and the emphasis is on informality instead of angular lines.

Problems

● The drive isn't part of the garden design, and this makes the area left for plants and grass look even smaller.
● The soil close to the base of a hedge is often dry and impoverished, so bedding plants don't thrive.

Solutions

● Most of the lawn has been dispensed with, and the flower beds enlarged and planted with low-maintenance shrubs. Plenty of evergreens have been used to provide year-round interest.
● Gravel has been used for the drive, and extended to form a broad and informal sweep to the front door. Not everyone likes gravel as a surface to walk on, however, and pavers could have been used instead. If plenty of plants cascade over the edge, the widening sweep would still look soft and attractive.

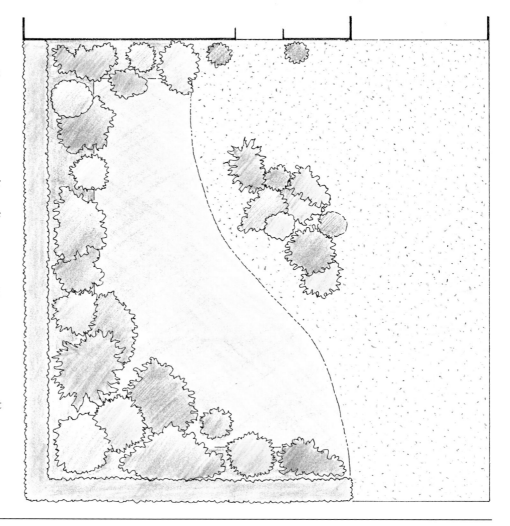

EXAMPLE TWO

Tall conifers along the drive dominate the garden and will continue to do so even after redesigning it. Remove trees that are too large rather than attempt to design around them.

Problems

● Tall hedges offer privacy, but here the scale is out of proportion, and depending on the aspect may keep out too much light.
● Rose beds are popular, but the small circular bed in the lawn looks incongruous with the rectangular design and can be difficult to mow around.
● Narrow, straight-edged beds around the edges make the lawn seem even smaller.

Solutions

● The concrete drive has been paved with bricks or brick-like pavers.
● A central planting strip has been left to break up the expanse of paving.
● The tall, dark hedge has been replaced with an attractive white ranch fence. The gravelled area beneath is planted with alpines.
● Climbing roses are planted by the house walls instead of a central weeping rose. They provide a fragrant welcome in summer.
● Existing borders remain to minimize the reconstruction.
● Small shrubs such as hebes and lavenders have been used, along with low-growing perennials like *Stachys lanata* (syn. *S. byzantina* or *S. olympica*) and *Bergenia cordifolia*, instead of seasonal bedding plants.
● A small deciduous tree, a crab apple, replaces the large conifer in the bottom corner. The area beneath can be planted with spring-flowering bulbs such as crocuses and snowdrops.
● The small circular bed has been enlarged and filled with gravel, as a base on which to stand pots.
● The narrow bed has been filled in with grass removed when the new paving in front of the house was laid. Bricks or blocks form a crisp edge.

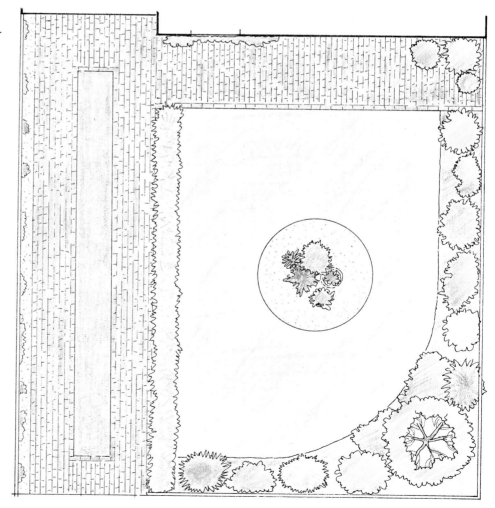

FRONT GARDENS

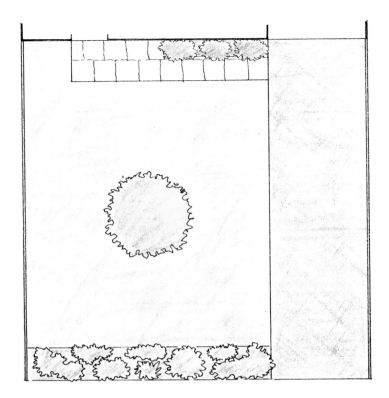

EXAMPLE THREE

Gardens don't come much more boring than this: a concrete drive, small narrow flower bed in front of the window and along the edge of the garden, and a single flowering cherry tree.

The solution for this garden was a simple one, as the redesigned garden shows. The cottage-garden style includes plants of all kinds which grow and mingle happily together with minimum intervention.

Besides being a short cut to the front door, the stepping stones encourage exploration of the garden and its plants. You actually walk through the planting, which cascades and tumbles around the paving slabs. The garden design has been reversed, with plants forming the heart of the garden rather than peripherals around the edge.

Problems

● Although the cherry is spectacular in flower, and provides a show of autumn colour, it is only attractive for a few weeks of the year. Its present position precludes any major redesign and so it is best removed.
● Unclothed wooden fences add to the drab appearance.
● Small flower beds like these lack impact, and are too small for the imaginative use of shrubs or herbaceous perennials.

Solutions

● The lawn and tree have been removed, and the whole area planted with a mixture of dwarf shrubs, herbaceous perennials, hardy annuals, and lots of bulbs for spring interest.
● Stepping-stones have been provided for those who want to take a short-cut (they also make access for weeding easier).
● The fences have been replaced with low walls so that the garden seems less confined.

EXAMPLE FOUR

This garden is a jumble of shapes and angles, and lacks any sense of design. With its new look, the old curved path has been retained because its thick concrete base and the drain inspection covers within it would have made it difficult to move, but all the other lines have been simplified and more appropriate plants used.

Problems

● Rock gardens are seldom successful on a flat site, and although small rock beds in a lawn can be made to resemble a natural rock outcrop, in this position the rocks can never look convincing.

● The tree here is young but will grow large and eventually cast considerable shade and dominate the garden.

● Small beds like this, used for seasonal bedding, are colourful in summer but can lack interest in winter. This curve sits uneasily with the straight edge at one end and the curve of the path at the other.

Solutions

● The rock garden has been paved so that the cultivated area is not separated by the drive.

● Gravel replaces the lawn. This needs minimal maintenance and acts as a good foil for the plants.

● Dwarf and medium-sized conifers create height and cover. By using species and varieties in many shades of green and gold, and choosing a range of shapes, this part of the garden now looks interesting throughout the year.

● Stepping-stones add further interest. Because it isn't possible to see where the stepping stones lead to from either end (the conifers hide the route), a sense of mystery is added and this tempts the visitor to explore.

● The existing path has been retained but covered with slate crazy-paving it looks more interesting.

● A pond creates a water feature.

● The awkward, narrow curving strip has been turned into a 'stream', with circulating water flowing over a cascade into the pond at one end.

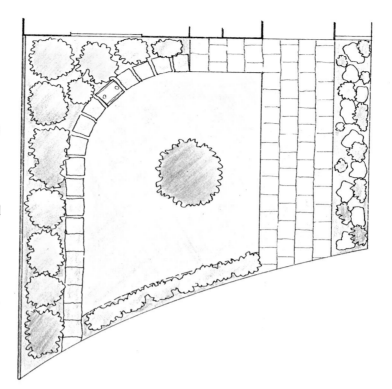

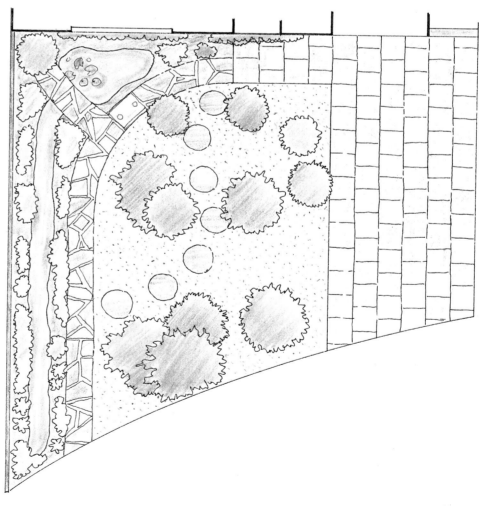

BASEMENT GARDENS AND BACKYARDS

Some gardens are not just small, they are gloomy too because they are below street level, or hemmed in by tall walls. Because there is little that can be done to alter this sort of garden structurally, it is best to direct any efforts towards improving the environment and devising a strategy that helps plants survive, or at least ensure lots of lush-looking plants to flourish despite the handicaps. Not all of the techniques shown here will be applicable to your own garden, but most of them can be adapted to suit even the most unpromising site.

Using lighting

Garden lights can extend the hours of enjoyment you derive from your garden, and you don't need many of them for a lot of impact in a small area. You can illuminate most of the space – useful if you often entertain in the evening – or use just one or two spotlights to pick out dramatic elements in the design. Some can be swivelled so that you can highlight different features. For subtle lighting, a cheaper and pretty option is to use lanterns which hold candles.

Painting the walls

In a garden enclosed by walls or fences, you need to do everything possible to reflect light and make the background bright and cheerful. Painting the walls a pale colour will improve things dramatically.

Using trellis

Trellis can be used as a decorative feature in its own right, or as a plant support. If you want to make a feature of it, paint it white, but if it is used primarily as a plant support, make sure it has been treated with a non-toxic preservative. Enclose unsightly downpipes in a trellis 'box' over which you can grow an evergreen climber such as ivy.

Adding water features

The sound of running water is refreshing on a summer's day, and in a small area you only need a trickle to do the job. A wall spout (with a tiny pool at ground level, from which the

ABOVE: *Ferns thrive in shady positions where many other plants would languish. If you can provide moisture from a water feature, so much the better.*
RIGHT: *Even the tiniest basement garden or backyard has space for a water feature.*

water is recirculated) or a self-contained wall fountain is ideal.

Introducing wind chimes

Wind chimes both look and sound good. Choose one primarily for the sound it makes.

Training wall shrubs

Cover some of the walls with climbers, but try espalier or fan-trained fruit trees or espalier pyracanthas too.

Furnishing in style

White-painted furniture looks bright in a small, enclosed garden, but don't add too much furniture or the area will look cluttered rather than elegant.

Using containers with character

If the area is small, make everything work for its space. Instead of plastic containers, use interesting old kitchen utensils, or other unexpected holders, but be sure to add drainage holes to prevent waterlogging.

Focal points in shade

Basement areas and enclosed backyards are often inhospitable for plants – the light is poor and the walls keep off much of the rain. If, in addition, you have a tree that casts shade, even the shade-loving plants will struggle. Use these positions for ornaments or make them into focal points.

Planting ferns

Ferns do well in a cool, shady spot, so use them freely in those areas too dull for bright summer flowers. Try a collection of hardy ferns – they won't look dull if you nestle an attractive ornament among them, or include white flowers, perhaps backed by a white wall. On a hot summer's day the space will be an oasis of coolness and tranquillity.

Growing white-flowered plants

Use pale flowers if the area lacks direct sun. You won't be able to use plants that need strong sun light, but fortunately some of the best white-flowering plants are shade-tolerant. Try white varieties of impatiens and white nicotianas, for example. White flowers will show up more brilliantly than coloured ones in a dull spot.

Introducing exotics

Gardens enclosed by walls can be hot and sunny too, and being sheltered provides the ideal environment for many exotic plants to grow successfully. Try a few bold houseplants to create a tropical effect.

LEFT: *Use wall pots and half baskets to make a dominant wall more interesting. They will be more effective staggered rather than in straight rows.*

LEFT BELOW: *White flowers, like this nicotiana, show up well in darker corners.*

Making the most of steps

Open railings can be used as supports for attractive climbers, planted in pots at the base of the steps, but always keep them trimmed so that slippery leaves do not trail across the steps or obstruct the hand-rail. If the steps are very wide, place pots of bright flowers on the steps themselves to produce a ribbon of colour. Do not obstruct the steps. If there is no space on the steps, use a group of containers filled with flowers at the top and bottom of the stairway.

Fixing windowboxes and wall baskets

Use windowboxes lavishly – not only beneath windows but fixed to walls too. Windowboxes, wall pots and half-baskets can all bring cascades of colour to a bare wall. Stagger the rows instead of placing them in neat and tidy lines.

Capturing the scents

An enclosed garden is an ideal place in which to grow scented plants – the fragrances are held in the air instead of being carried off on the wind. Use plenty of aromatic plants, especially big and bold plants like daturas, and those with a heavy perfume such as evening-scented nicotianas and night-scented stocks.

BALCONIES AND VERANDAS

For someone without a garden, a balcony may be their entire 'outdoor room', a 'garden' to enjoy from indoors when the weather is inclement. Even more than a patio, the balcony or veranda is an outdoor extension of the home.

The area is usually small, so the money you are prepared to spend on gardening will go a long way. Splash out on quality flooring and furniture, and ornate containers, which will create a classy setting for your plants.

Choosing flooring

The floor will help to set the tone and style, and it can make or mar a tiny 'garden' like this.

Paving slabs are best avoided: they are heavy, frequently lack the kind of refinement that you can achieve with tiles, and the size of individual slabs may be too large to look 'in scale' for the small area being covered.

Think of the veranda or balcony floor as you might the kitchen or conservatory floor – and use materials that you might use indoors. Quarry tiles and decorative ceramic tiles work well, and produce a good visual link with the house. Make sure ceramic tiles are frostproof however. Tiles are relatively light in weight, and their small size is in proportion to the area.

Timber decking is another good choice for a veranda.

The problem of aspect

Aspect is an important consideration. Unlike a normal garden, or even a roof garden, the light may be strong and intense all day, or there may be constant shade, depending on position. Balconies above may also cast shade.

If the aspect is sunny, some shade from above can be helpful. Consider installing an adjustable awning that you can pull down to provide shade for a hot spot. Choose sun-loving plants adapted to dry conditions for this situation – your indoor cacti and

RIGHT: *Roof gardens and balconies are often improved if you lay a wooden floor and create a timber overhead.*

succulents will be happy to go outside for the summer.

If the aspect is shady for most of the day a lot of flowering plants won't thrive. You may have to concentrate on foliage plants, though some bright flowers, such as impatiens and nicotiana, do well in shade.

Countering the wind

Like roof gardens, balconies are often exposed to cold and damaging winds. The higher a balcony the greater problem wind is likely to be.

To grow tender and exotic plants, provide a screen that will filter the wind without causing turbulent eddies. A trellis clothed with a tough evergreen such as an ivy is useful, or use screens of woven bamboo or reeds on the windiest side – these not only provide useful shelter and privacy, but make an attractive backdrop for plants in containers.

Adding colour round the year

Create a framework of tough evergreens to clothe the balcony or veranda throughout the year, and provide a backdrop for the more colourful seasonal flowers.

Use plenty of bright seasonal flowers in windowboxes or troughs along the edge, with trailers that cascade down over the edge.

In the more sheltered positions, grow lots of exotic-looking plants, and don't be afraid to give lots of your tough-leaved houseplants a summer holiday outside.

Pots of spring-flowering bulbs extend the season of bright flowers, but choose compact varieties – tall daffodils, for example, will almost certainly be bent forward as wind bounces back off the walls.

Add splashes of colour with cut flowers. In summer choose long-lasting 'exotics' such as strelitzias and anthuriums.

ABOVE: *In mild areas or a sheltered position, you can turn your balcony into a tropical garden.*

RIGHT: *Turn your balcony into an outdoor room where many indoor plants thrive in summer.*

FEATURES
and STRUCTURES

Overall garden design is important, but it is individual features that make a garden special. Major structural decisions, such as the type of paving to use, the shape of the lawn, or how to define the boundaries, have a significant impact, but even small details like ornaments and garden lights can lift a small garden above the ordinary. The use of containers is especially important in a small garden – on a tiny balcony they may be the garden. Use them imaginatively, choosing containers that are decorative, and grouping them for added interest.

ABOVE: *Create the urge to explore with small paths that lead to features such as seats and ornaments.*

OPPOSITE: *The garden floor is important, whether paving or a lawn, but it is features, like this arbour and its seat, that give the garden character.*

the GARDEN FLOOR

THE GARDEN FLOOR – LAWN, PAVING, PATHS, even areas of gravel or ground cover plants – can make or mar your garden. These surfaces are likely to account for more area than the beds and borders. Although they recede in importance when the garden is in full bloom, for much of the year they probably hold centre stage.

Removing existing paths and paved areas presents a practical problem. If they are laid on a thick bed of concrete you will probably have to hire equipment to break up the surface. Provided these areas do not compromise your design too much, it is much easier to leave as many as you can in position. Consider paving over the top with a more sympathetic material. It should be relatively easy to extend the area if you want to.

Lawns are more easily modified than paths and paved areas. At worst you can dig them up and resow or relay them. If you simply want to change the shape, you can trim off surplus grass or lift and relay just part of the lawn.

ABOVE: *Paths can be both functional and attractive, often giving the garden shape and form.*

OPPOSITE: *Hard landscaping, such as bricks, combined with soft landscaping, such as lawns, can look very harmonious if designed with integration in mind.*

OPPOSITE ABOVE: *A brick edging marks the boundary between lawn and border, and serves the practical purpose of making mowing easier.*

LEFT: *Areas like this would soon become weedy if not densely planted. Here hostas suppress the weeds, and Soleirolia soleirolii spills over onto the path.*

Timber decking is very popular in some countries, seldom used in others. Much depends on the price of timber locally, and to some extent the climate, but decking should always be on your list of options.

There are useful alternatives to grass for areas that are not used for recreation or are seldom trodden on. Ground cover plants not only suppress weeds in flower beds, but can replace a lawn where the surface does not have to take the wear and tear of trampling feet. Inset stepping-stones to protect the plants. Where the garden is *very* small, low-growing ground cover may be much more practical than a lawn that is almost too tiny to cut with a mower.

LAWNS

The lawn is often the centrepiece of a small garden, the canvas against which the rest of the garden is painted. For many gardeners this makes it worth all the mowing, feeding and grooming that a good lawn demands. If your lawn has to serve as a play area too, be realistic and sow tough grasses, and settle for a hard-wearing lawn rather than a showpiece. It can still look green and lush – the important consideration from a design viewpoint. Instead of aiming for a bowling-green finish, the shape of the lawn or a striking edging could be its strong visual message.

Working with circles

Circular lawns can be very effective. Several circular lawns, linked by areas of paving, such as cobbles, work well in a long, narrow garden.

If the garden is very small, all you will have space for is a single circular lawn. If you make it the centrepoint with beds around it that become deeper towards the corner of the garden, you will be able to combine small trees and tall shrubs at the back with smaller shrubs and herbaceous plants in front. To add interest, include a couple of stepping-stone paths that lead to a hidden corner.

Using rectangles

Rectangular lawns can look boring, but sometimes they can be made more interesting by extending another garden feature – such as a patio or flower bed – into them to produce an L-shaped lawn.

Alternatively, include an interesting feature such as a birdbath or sundial (often better towards one side or end of the lawn than in the middle). A water feature is another good way to break up a boring rectangle of grass.

An angled lawn

If you have chosen a diagonal theme for your design, you will probably want to set your lawn at an angle to the house so that it fits in with the

ABOVE RIGHT: *A sweeping lawn can help to create a sense of perspective.*
RIGHT: *This lawn would look boring with straight edges. The curves add style.*

other features. The same rectangle of lawn becomes much more interesting when set at an angle of about 45 degrees. By lifting and patching the lawn, you may be able to achieve this without having to start from scratch.

Creating curves

A sweeping lawn with bays and curves where the flower borders ebb and flow is very attractive. It is difficult to achieve in a small garden. However, you can bring out a border in a large curve so the grass disappears around the back. You may be able to do this by extending the border into an existing rectangular lawn.

Changing height

If you have to create an impression in a small space, try a raised or sunken lawn. The step does not have to be large – 15–23cm (6–9in) is often enough. If making a sunken lawn, always include a mowing edge so that you can use the mower right up to the edge of the grass.

ABOVE: *Sunken lawns make a bold feature.*

KEEPING A TRIM EDGE

Circular lawns must be edged properly. Nothing looks worse than a circle that isn't circular, and of course constant trimming back will eat into the lawn over the years. To avoid this, incorporate a firm edging, such as bricks placed on end and mortared into position, when you make the lawn.

Where the edges are straight use proprietary lawn edging strips.

HOW TO CREATE A MOWING EDGE

If flowers tumble out of your borders, or there is a steep edge that makes mowing difficult, lay a mowing edge of bricks or paving slabs.

1 Mark out the area of grass to be lifted using the paving as a guide. Lift the grass where you want to lay the paved edge. To keep the new edge straight, use a half–moon edger against the paving slab. Then lift the grass to be removed by slicing it off with a spade.

2 Make a firm base by compacting gravel or a mixture of sand and gravel where the paving is to be laid. Use a plank of wood to ensure it is level. Allow for the thickness of the paving and a few blobs of mortar.

3 It is best to bed the edging on mortar for stability, but as it will not be taking a heavy weight just press the slabs onto blobs of mortar and tap level (use a spirit–level to double–check).

IMAGINATIVE PAVING

Most small gardens have a patio or at least a paved area close to the house. Often it is the main feature around which the remainder of the back garden is arranged. It can be the link that integrates home and garden. At its worst, paving can be boring and off-putting; at its best it can make a real contribution to the overall impact of the garden.

On the following pages you will find a selection of popular paving materials, with suggestions for use, and their advantanges and disadvantages. Always shop around because the availability and price of natural stones vary enormously, not only from country to country, but also from area to area.

Even the availability of man-made paving will vary from one area to another. Choosing the material is only part of the secret of successful paving – how you use it, alone or combined with other materials, is what can make an area of paving mundane or something special.

Colour combinations
Your liking for bright and brash colour combinations will depend on the effect you want to create. Be wary of bright colours though – they can detract from the plants, although they will mellow with age.

Sizing up the problem
In a small garden, large-sized paving units can destroy the sense of scale. Try small-sized paving slabs (which are also easier to handle), or go for bricks, pavers, or cobbles.

Mix and match
Mixing different paving materials can work well, even in a small space. Try areas or rows of bricks or clay pavers with paving slabs, railway sleepers with bricks, in fact any combination that looks good together and blends with the setting. Avoid using more than three different materials, however, as this can look too fussy in a small garden.

LEFT: *Bricks and pavers often look more attractive if laid to a pattern such as this herringbone style.*

Paving patterns
You can go for a completely random pattern – crazy-paving is a perfect example – but most paving is laid to a pre-planned pattern using rectangular paving slabs or bricks. Look at the brochures for paving slabs. These usually suggest a variety of ways in which the slabs can be laid.

Although a large area laid with slabs of the same size can look boring, avoid too many different sizes, or complex patterns in a small space. Simplicity is often more effective.

Bricks and clay pavers are often the best choice for a small area, because their small size is more likely to be in harmony with the scale of the garden. The way they are laid makes a significant visual difference, however, so choose carefully.

The stretcher bond is usually most effective for a small area, and for paths. The herringbone pattern is suitable for both large and small areas, but the basket weave needs a reasonably large expanse for the pattern to be appreciated.

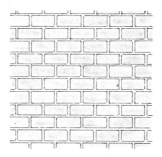

Stretcher bond

Herringbone

Basket weave

HOW TO LAY PAVING

1 Excavate the area to a depth that will allow for about 5cm (2in) of compacted hardcore topped with about 3–5cm (1–2in) of ballast, plus the thickness of the paving and mortar. As an alternative to hardcore topped with ballast, you can use 5cm (2in) of scalpings. Check the depth of the foundation before laying the paving. If adjoining the house, make sure that the paving will end up below the damp-proof course.

2 Put five blobs of mortar where the slab is to be placed – one at each corner, and the other in the middle.

3 Alternatively, cover the area where the paving is to be laid with mortar, then level.

4 Position the slab carefully, bedding it on the mortar.

5 Use a spirit-level to ensure that the slab is level, but use a small wedge of wood under one end to create a slight slope over a large area of paving so that rainwater runs off freely. Tap the slab down further, or raise it by lifting and packing in a little more mortar. Position the level over more than one slab (place it on a straight-edged piece of wood if necessary).

6 Use spacers of an even thickness to ensure regular spacing. Remove these later, before the joints are filled with mortar.

7 A day or two after laying the paving, go over it again to fill in the joints. Use a small pointing trowel and a dryish mortar mix to do this. Finish off with a smooth stroke that leaves the mortar slightly recessed. This produces an attractive, crisp look. Wash any surplus mortar off the slabs before it dries.

PAVING MATERIALS

There are plenty of paving materials from which to choose, so spend time looking through brochures and visit garden centres and builders' merchants before you come to a decision.

RIGHT: *Bricks, unlike clay pavers, are laid with mortared joints. This can emphasize the design.*

PAVING SLABS

Rectangular paving slabs

The majority of paving slabs are based on a full-sized slab 45 × 45cm (18 × 18in) or 45 × 60cm (18 × 24in). Half and quarter slabs may be a little smaller in proportion to allow for mortar joints. Thickness may vary according to make, but provided you mix only those made by the same manufacturer this won't matter.

A *smooth* surface can be boring, slippery, and a little too much like public paving, but many have a *textured* finish. Textures vary. A riven finish usually looks like natural stone, an exposed aggregate finish has exposed gravel to give a natural-looking non-slip finish.

Slabs imprinted with a section of a larger pattern are usually unsatisfactory in a small area. As quite a large area of paving is usually required to complete the pattern, they only emphasize the space limitations.

Shaped paving slabs

Use shaped slabs with caution. Circular slabs are useful for stepping-stones, but are difficult to design into a small patio. Hexagonal slabs also need a fairly large area to be appreciated. Special half-block edging pieces are usually available to produce a straight edge.

Paved and cobbled finish slabs

Some designs are stamped with an impression to resemble groups of pavers or bricks, some containing as many as eight basket-weave 'bricks' within the one slab. They create the illusion of smaller paving units, and are very effective in a small area.

TOP LEFT: *Slabs like this are particularly useful for a small area because they give the illusion of smaller paving units.*

TOP RIGHT: *Paving slabs with a riven finish look convincingly like real weathered stone.*

MIDDLE LEFT: *Paving slabs will always weather. Pale colours like this will soon look darker, while bright colours will become muted.*

MIDDLE RIGHT: *Hexagonal paving slabs can be attractive, but are not usually satisfactory in a very small area.*

BOTTOM: *Rectangular shapes like this can be used alone, or integrated with other sizes to build up an attractive design.*

Planting circles

A few manufacturers produce paving slabs with an arc taken out of one corner. Four of these placed together leave a circular planting area for a tree or other specimen plant.

BRICKS AND PAVERS

Bricks and pavers are especially useful for a small garden. You can create an attractive design even in a small area, and you may be able to obtain them in a colour and finish that matches your home, which will produce a more integrated effect.

Always check that the bricks are suitable for paving, however, as some intended for house building will not withstand the frequent saturation and freezing that paths and patios are subjected to. After a few seasons they will begin to crumble. Clay pavers, on the other hand, have been fired in a way that makes them suitable for paving. Concrete pavers and blocks are another option, though these are usually more suitable for a drive than a small patio.

Rectangular pavers

Clay pavers look superficially like bricks but are designed to lock together without mortar. They are also thinner than most bricks, though this is not obvious once they have been laid. Concrete pavers or paving blocks are laid in a similar way and are more attractive than concrete laid in-situ for a drive. They can look a little 'municipal'.

Interlocking pavers and blocks

Concrete pavers or blocks are often shaped so that they interlock. Interlocking clay pavers may also be available.

Bricks

Bricks require mortar joints – they won't interlock snugly like clay pavers. On the other hand you may be able to use the same bricks for raised beds and low walls, giving the whole design a more planned and well-integrated appearance.

To use bricks economically, lay them with their largest surface exposed, not on edge. This excludes the use of pierced bricks (which have holes through them). It does not matter if they have a frog (depression) on one side, provided this is placed face-down.

Setts and cobbles

Imitation granite setts, which are made from reconstituted stone, and cobbles, which are natural, large, rounded stones shaped by the sea or glaciers, are both excellent for small areas of irregular shape. Their size makes them much easier to lay to a curve. Bed them into a mortar mix on a firm base.

Tiles

Quarry and ceramic tiles are appropriate for small areas near the house, or to create a patio that looks just that little bit different. Always make sure ceramic tiles are frostproof. Lay them on a concrete base that has been allowed to set, and fix them with an adhesive recommended by the supplier or manufacturer.

LEFT: *Hard paving comes in many forms. The top row shows (from left to right) natural stone sett, clay paver, clay brick, artificial sett. The centre row shows a typical range of concrete paving blocks. The bottom row illustrates some of the colours available in concrete paving slabs.*

PATHS AND PATH MATERIALS

As with any other garden structure, paths should be designed to suit the purpose they are to serve. There are a wide range of materials on the market to suit every need so shop around before deciding which you require.

Practical paths should be functional first and attractive second. Drives for cars and paths to the front door must be firmly laid on proper foundations. And don't skimp on width – it is extremely frustrating for visitors if they have to approach your door in single file. It might be better for the route to take a detour, perhaps forming an L-shape with the drive, if there isn't enough space for a wide path directly to the door.

Internal paths, used to connect one part of the garden to another, can be more lightly constructed, and are softened with plants.

Casual paths, which often lead nowhere and are created for effect, such as stepping-stones through a flower bed, can be lightly constructed and much less formal in style.

Bricks and pavers

These are ideal materials for internal garden paths that have to be both practical and pretty. Complex bonding patterns are best avoided unless the path is very wide.

Paving slabs

By mixing them with other materials the look of paving slabs can be much improved. A narrow gravel strip either side can look smart, and the gravel can be extended between the joints to space out the slabs. The slab-and-gravel combination is ideal if you need a curved path.

A straight path can be broken up with strips of beach pebbles mortared between the slabs. Tamp them in so that they are flush with the surrounding paving.

RIGHT: *Although the gaps between these paving slabs have been filled with chipped bark in this example, you could also use gravel.*

BELOW: *Paving can reflect artistic ambitions.*
BELOW RIGHT: *Victorian-style rope edging.*

Crazy-paving

Use this with caution. In the right place, and using a natural stone, the effect can be mellow, and harmonize well with the plants. Be more wary of using broken paving slabs – even though they are cheap. Coloured ones can look garish, and even neutral slabs still look angular and lack the softness of natural stone.

Path edgings

Paths always make a smarter feature with a neat or interesting edging. If you have an older-style property, try a Victorian-style edging. If it is a country cottage, try something both subtle and unusual, like green glass bottles sunk into the ground so that just the bottoms are visible. Or use bricks: on their sides, on end, or set at an angle of about 45 degrees.

CREVICE PLANTS

Plants look attractive and soften the harsh outline of a rigid or straight path. They are easy to use with crazy-paving or any path edged with gravel. It may be necessary to excavate small holes. Fill them with a good potting mixture. Sow or plant into these prepared pockets.

Some of the best plants to use for areas likely to be trodden on are chamomile, *Thymus serpyllum* and *Cotula squalida*. For areas not likely to be trodden there are many more good candidates, such as *Ajuga reptans* and *Armeria maritima*.

HOW TO LAY CLAY OR CONCRETE PAVERS

The method of laying clay or concrete pavers described in the following steps can be used for a drive or a patio as well as a path.

1 Excavate the area and prepare a sub-base of about 5cm (2in) of compacted hardcore or sand and gravel mix. Set an edging along one end and side first, mortaring into position, before laying the pavers.

2 Lay a 5cm (2in) bed of sharp sand over the area, then use a straight-edged piece of wood stretched between two height gauges (battens fixed at the height of the sand bed) to strike off surplus sand and provide a level surface.

3 Position the pavers, laying 2m (6½ft) at a time. Make sure they butt up to each other, and are firm against the edging. Mortar further edging strips into place as you proceed.

4 Hire a flat-plate vibrator to consolidate the sand. Alternatively, tamp the pavers down with a club hammer over a piece of wood. Do not go too close to an unsupported edge with the vibrator.

5 Brush more sand into the joints, then vibrate or tamp again. It may be necessary to repeat this once more.

TIMBER DECKING

Timber decking creates a distinctive effect, and will make a refreshing change from ordinary paving for the patio area. As with paving, the material used should be in proportion to the size of the garden, so the width of the planks is important. Wide planks look best in a large garden, but in a small, enclosed area narrower planks are usually preferable.

Different designs can be achieved by using planks of different widths and fixing them in different directions, as illustrated here, but on the whole it is best to keep any pattern fairly simple. Leave a small gap between each plank, but not so large that high-heeled shoes can slip into it.

The construction method and timber sizes must reflect the size of the overall structure and its design – especially if built up over sloping ground. In some countries there are building codes and regulations that may have to be met. If in doubt, seek professional help with the design, even if you construct it yourself.

All timber used for decking must be thoroughly treated with a wood preservative. Some preservatives and wood stains are available in a range of colours, and this provides the opportunity for a little creativity. Dark browns and black always look good and weather well, but if you want to be more adventurous choose from reds, greens and greys.

If you want your decking to have a long life, special pressure-treated timber is the best choice. However, the range of colours available is bound to be less extensive.

Parquet decking

The easiest way to use timber as a surface is to make or buy parquet decking. Provided the ground is flat panels are easy to lay and can look very pleasing. Bed them on about 5cm (2in) of sand over a layer of gravel, to ensure free drainage beneath. If you already have a suitable concrete base to use, you can lay them directly onto this.

LEFT: *Timber decking makes a refreshing change from paving slabs or bricks, and can give the garden a touch of class.*

Patterns of Timber Decking

 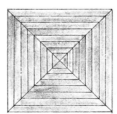 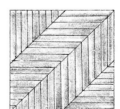 

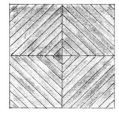   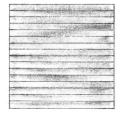

GROUND COVER WITH PLANTS

If you want to cover an area of ground with a living carpet simply for texture, and don't expect to walk on the area, suitable ground cover plants are the answer.

To use ground cover plants like this, rather than simply as a means to suppress weeds in a flower bed, they must be evergreen, compact, and grow to a low, even height.

HOW TO PLANT CLUMP-FORMING GROUND COVER

1 Clear the ground of weeds first, and be especially careful to remove any deep-rooted or difficult perennial weeds.

2 Add plenty of garden compost or rotted manure, then rake in a controlled-release fertilizer. Add these before laying a mulching sheet.

3 Cover the area with a weed-suppressing mulching sheet. You can use a polythene sheet, but a special woven mulching sheet is much better.

4 Make crossed slits through the sheet where you want to plant. Avoid making the slits too large.

5 Excavate planting holes and firm in the plants. If necessary tease a few of the roots apart first.

6 Water thoroughly, and keep well watered. Remove the sheet once the plants are well established.

GROUND COVER PLANTS

Some of the best plants for the job are *Armeria maritima*, bergenias, *Cotoneaster dammeri*, *Euonymus fortunei* varieties, *Hypericum calycinum*, and *Pachysandra terminalis*. If you want flowers as the main feature, heathers are difficult to better.

HOW TO PLANT CREEPING GROUND COVER

The mulching sheet method is a good way to get clump-forming plants such as heathers off to a good start, but don't use it for those that creep and root, such as ajugas and *Hypericum calycinum*. Plant these normally but apply a loose mulch about 5cm (2in) thick to cover the soil.

GRAVEL GARDENS

Gravel is an inexpensive and flexible alternative to paving or a lawn, although it is not suitable for a patio. It blends beautifully with plants, needs little maintenance, and can be used in both formal and informal designs. It is also a useful 'filler' material to use among other hard surfaces, or in irregularly shaped areas where paving will not easily fit and a lawn would be difficult to mow.

LEFT: *Gravels naturally vary considerably in colour.*

Types of gravel

Gravel comes in many different shapes, sizes and colours. Some types are angular, others rounded, some are white, others assorted shades of green or red. All of them will look different in sun or shade, when wet or dry. The subtle change of colour and mood is one of the appeals of gravel. The gravels available will depend on where you live, and which ones can be transported economically from further afield. Shop around first going to garden centres and builders merchants to see what is available in your area before making your choice.

Gravel paths

Gravel is often used for drives, but it is also a good choice for informal paths within the garden. It conforms to any shape so is useful for paths that meander. However, it is not a good choice for paths where you will have to wheel the mower.

HOW TO LAY A GRAVEL PATH

1 Excavate the area to a depth of about 15cm (6in), and ram the base firm.

2 Provide a stout edge to retain the gravel. For a straight path, battens secured by pegs about 1m (3ft) apart is an easy and inexpensive method.

3 First place a layer of compacted hardcore. Add a mixture of sand and coarse gravel (you can use sand and gravel mixture sold as ballast). Rake level and tamp or roll until firm.

4 Top up to the required height with the final grade of gravel. In small gardens, the size often known as pea gravel looks good and is easy to walk on. Rake and roll repeatedly until the surface is firm and stable.

If the path is wide, it is a good idea to build the gravel up towards the centre slightly so that puddles do not form after heavy rain.

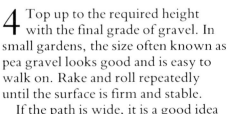

Gravel beds

Gravel can be used as a straight substitute for grass and requires much less maintenance. You can even convert an existing lawn very simply by applying a weedkiller to the grass, laying edging blocks around the edge, then topping up with gravel.

Informal gravel beds still require some kind of edging restraint to prevent the gravel from spreading. If the bed is surrounded by a lawn, simply make sure that the gravelled area is about 5cm (2in) below the surrounding grass.

Other practical ways to prevent the gravel from scattering onto beds and other unwanted areas are to create a slightly sunken garden or to raise the surround slightly with a suitable edging.

Informal gravel areas often look especially effective if some plants are grown through the gravel – either in beds with seamless edges where the gravel goes over them, or as individual specimen plants.

HOW TO LAY A GRAVEL BED

1 Excavate the area to the required depth – about 5cm (2in) of gravel is sufficient in most cases.

2 Level the ground. Lay heavy-duty black polythene or a mulching sheet over the area. Overlap strips by about 5cm (2in).

3 Then tip the gravel on top and rake level.

4 To plant through the gravel, draw it back from the planting area and make a slit in the polythene. Plant normally, enriching the soil beneath if necessary.

5 Firm in and pull back the polythene before re-covering with gravel.

FORMING BOUNDARIES

Most of us have an instinctive desire to mark our territory with a very visible boundary. It gives us a sense of privacy and the illusion of security, but above all it marks out our plot of land, the area in which we create our own very special paradise.

The problem with a small garden is that the boundary forms a large part of the garden, and the chances are that you will see it from whichever direction you look. In a large garden the boundary often merges into the background, but in a small one it can easily dominate.

Tall walls can be an asset – the walled town garden has many of the treasured attributes of an old walled country garden – but drab wooden fences and large overgrown hedges pose real problems if you want to make your garden look smart and stylish.

Don't take your boundary for granted, and never assume it can't be improved. Replacing a fence or grubbing up a long-established hedge are not projects to be tackled lightly – they can be expensive or labour-intensive. Never make changes until you have consulted neighbours that

LEFT: *This is an excellent example of a combination boundary – a wooden picket fence supported on a low wall, with an escallonia flowering hedge growing through it.*

OPPOSITE ABOVE: *Walls make secure boundaries, but to prevent them looking oppressive cover with climbers, and if possible create a view beyond, as this attractive gate has done.*

OPPOSITE: *A wall as tall as this can easily dominate a small garden, but by treating it boldly and using it as a feature it becomes an asset.*

are affected. The boundary may belong to them, in which case it is not yours to change unilaterally. Even if it is legally yours to replace, the courtesy of discussing changes with others affected will go a long way to helping you remain on good terms with your neighbours.

Although you are unlikely to want to exceed them in a small garden, there may be legal limitations on boundary height, perhaps laid down in the terms of the contract when you bought the property. In some countries there may be restrictions placed by the highways authority on road safety grounds.

Restrictions are most likely in front gardens – some 'open plan' estates, for example, may have limitations on anything that might infringe the integrated structure of the gardens.

None of these restrictions need inhibit good garden design, but it is always worth checking whether any restrictions exist before erecting or planting a new boundary.

HEDGES FOR SMALL GARDENS

Many of the classic hedges, like beech, yew, and tall conifers like × *Cupressocyparis leylandii*, and even the privet (*Ligustrum ovalifolium*) have strictly limited use in a small garden. In small gardens the emphasis should be on plants that have much to offer or compact growth. The hedges suggested here are just some of the plants that could be used to mark your boundary without being dull or oppressive. Be prepared to experiment with others.

Clipped formality

The classic box hedge (*Buxus sempervirens*) is still one of the best. It clips well and can be kept compact, but choose the variety 'Suffruticosa' if you want a really dwarf hedge like those seen in knot gardens. A quick-growing substitute is *Lonicera nitida*, and there's a golden form that always looks bright – but be prepared to cut frequently. Some of the dwarf berberis stand close clipping – try the red-leaved *Berberis thunbergii* 'Atropurpurea Nana'. Yew (*Taxus baccata*) is also excellent for formal clipping, and it can be kept compact enough for a small garden.

Colourful informal hedges

If you want to cut down on clipping, and want something brighter and more colourful than most foliage hedges, try the grey-leaved *Senecio* 'Sunshine' or the golden *Philadelphus coronarius* 'Aureus' (unfortunately sheds its leaves in winter). *Viburnum tinus* can also be kept to a reasonable height, and provided you avoid pruning out the new flowers it will bloom in winter. Many of the flowering and foliage berberis also make good 'shrubby' hedges. These will lack a neatly clipped profile, but pruning and shaping is normally only an annual job.

ABOVE: *Although* Lonicera nitida *needs frequent clipping, it makes a neat formal hedge.*

LEFT: *Many shrub roses can make an attractive flowering hedge in summer, but do not plant them too close to the edge of a path otherwise their thorny stems may be a nuisance.*

Using roses

Roses make delightful – and often fragrant – boundaries, but they have shortcomings. Their summer beauty is matched by winter ugliness, and they are not a good choice for a boundary where passers-by may be scratched by thorns. You can use a row of floribunda (cluster-flowered) roses, but the shrubby type are usually preferred for this job.

Old-fashioned lavender and rosemary dividers

Both these herbs make excellent informal flowering hedges, with the merit of being evergreen too. You could try the shorter lavender in front of the taller rosemary. Both become untidy with age, so replace the plants when it becomes necessary.

Other flowering hedges

Forsythia is one of the most popular flowering hedges, but careful pruning is required to achieve consistent flowering on a compact hedge. There are plenty of alternatives, including the shrubby potentillas, berberis like B. × *stenophylla*, with bold flowers, though this one can take up a lot of space, and even tall varieties of heathers if you just want a boundary marker rather than a barrier.

ABOVE: *Box is one of the classic plants to use for clipped formality. This is a glaucous form.*

HOW TO PLANT A NEW HEDGE

1 Prepare the ground very thoroughly. Excavate a trench – ideally about 60cm (2ft) wide – and fork in plenty of rotten manure or garden compost.

2 Add a balanced fertilizer at the rate recommended by the manufacturer. Use a controlled-release fertilizer if planting in the autumn.

3 Use a garden line, stretched along the centre of the trench, as a positioning guide. If the area is windy or you need a particularly dense hedge, plant a double row of trees. Bare-root plants are cheaper than container-grown plants, but only separate them and expose the roots once you are ready to plant. Only the most popular hedging plants are likely to be available bare-root, and for many of the plants suggested you will have to plant a single row of container-grown plants.

4 Use a piece of wood or a cane cut to the appropriate length as a guide for even spacing. Make sure the roots are well spread out. If planting container-grown plants, tease out some of the roots that are running around the edge of the root-ball.

5 Firm the plants in well and water thoroughly. Be prepared to water the hedge regularly in dry weather for the first season. Keep down weeds until the hedge is well established, then it should suppress the weeds naturally.

GARDEN WALLS

Except for special cases, such as basement flats and the need for privacy or screening in a difficult neighbourhood, high walls are inappropriate for a small garden. However, low walls up to about 1–1.2m (3–4ft) are a useful alternative to a hedge, particularly if you want to avoid regular trimming. Although the rain shadow and shade problems remain the same for a wall as a hedge, a wall will not impoverish the soil in the same way as a hedge.

Low walls

A low wall, say 30–60cm (1–2ft) tall, will serve the same demarcation function as a taller one, but in more appropriate scale for a small garden, and shrubs planted behind it are more likely to thrive. Modest garden walls like this are much easier to construct than tall ones, which may need substantial reinforcing piers, and are well within the scope of a competent garden handyman to build.

Brick walls

Plain brick walls can harmonize with the house, but generally look dull from a design viewpoint. A skilled bricklayer can often add interest by laying panels or strips to a different pattern. The choice of brick and the capping will also alter the appearance. Some are capped with bricks, others have special coping tiles. These all add to the subtle variety of brick walls.

Block walls

Many manufacturers of concrete paving slabs also produce walling blocks made from the same material. These are especially useful for internal garden walls and raised beds. They are often coloured to resemble natural stone, but brighter colours are available if you want to match the colour scheme used for the paving. Bear in mind that colours will weather and become much more muted within a couple of years.

ABOVE: *An interesting brick wall.*
RIGHT: *A wall like this makes a solid and secure boundary without making the garden appear too enclosed.*

Screen block walls

Screen block walls (sometimes called pierced block) are most frequently used for internal walls, perhaps around the patio or to divide one part of the garden from another, but they can also be used to create a striking boundary wall too.

These blocks have to be used with special piers and topped with the appropriate coping. They are useful if you want to create a modern image, or perhaps the atmosphere of a Mediterranean garden.

Mixing materials

Some of the smartest boundary walls are made from more than one material. Screen blocks look good as panels within a concrete walling block framework. Screen blocks can also be incorporated into a brick wall, and help to let light through and to filter some of the wind. Panels of flint or other stone can be set into an otherwise boring tall brick wall.

BELOW: *Walls can be colourful . . . if you create planting areas. The summer bedding plants used here are replaced at the end of the season with bulbs and spring bedding plants to use the planting areas to their full advantage.*

Cavity walls

Low cavity walls that can be lined with plants soon become an eye-catching feature. Pack them with colourful summer flowers, or plant with permanent perennials such as dwarf conifers, which maintain interest throughout the year (but be sure to choose truly dwarf conifers for this). If you plant cascading forms

ABOVE: *Dry stone walls are not difficult to build provided you keep them low, and you can plant into the sides for extra interest.*

such as nasturtium or trailing lobelia, the effect can be really stunning. For a spring display, try aubrietia and the yellow *Alyssum saxatile*, with a few dwarf spring-flowering bulbs.

Dry stone walling

Dry stone walls are more often used for retaining banks or as internal dividers, but in an appropriate setting this kind of wall makes an attractive boundary. This type of wall looks best where dry stone walls are part of the natural landscape.

The great advantage of a dry stone wall – which is assembled without mortar – is the ability to plant in the sides. This can provide a home for many kinds of alpines.

Walls with a difference

The larger and taller the wall, the more imaginative you should be when designing it. Try incorporating an alcove for an ornament, or a panel into which you can set an artistic piece of wrought-iron that can be viewed against the green of a neighbouring field or garden.

WALL BUILDING MATERIALS

Most builders' merchants stock a good range of bricks, the majority are suitable for garden walls, but if you need a lot of bricks – enough to justify ordering direct – get in touch with a few brick companies. Their expertise can be invaluable, and most will be able to offer you a wide choice.

Buying bricks is something most of us do only rarely, so professional advice is especially useful. The author's experience, however, suggests that you can't always depend on the advice of a builder's merchant. Shop around until you find someone who really appears to have a knowledgeable passion for bricks – they will tell you about all the different finishes and colours available, and most importantly will know whether a particular brick is suitable for the job you have in mind. *Always* explain what you want your bricks for: a building, a garden wall, wall of a raised bed, or for paving. Some bricks which are perfectly suitable for house walls may be very unsuitable for paths or garden walls.

If you need a lot of bricks (many hundreds), it may be better, and cheaper, to buy direct from a brick manufacturer if they will deal with the general public.

ABOVE: *Bricks come in many colours and finishes, and these are just a small selection from the many kinds available. Names of bricks vary from country to country, but whatever names are used you are likely to have a good choice.*

MASONRY MORTAR

A suitable mortar for bricklaying can be made from 1 part cement to 3 parts soft sand. Parts are by volume and not weight. Cement dyes can be added to create special effects, but use coloured mortar cautiously.

Common bonds

Expert brick-layers may use more complicated bonds, but for ordinary garden walls – and especially those that you are likely to lay yourself, perhaps for a low boundary wall or for a raised bed within the garden – it is best to choose one of the three bonds illustrated below.

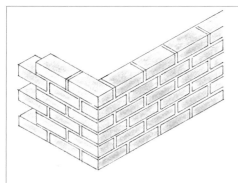

Running bond or stretcher bond
This is the simplest form of bonding, and is used for walls a single brick wide – or where you want to create a cavity, such as a low wall with a planting space.

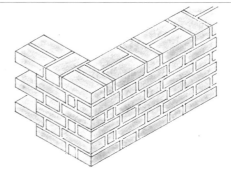

Flemish bond This is another way to create a strong bond in a wall two bricks wide. The bricks are laid both lengthways and across the wall within the same course.

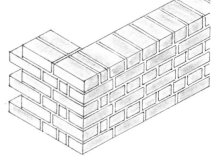

English bond This is used for a thick wall the width of two bricks laid side by side – useful where strength is needed for a high wall. Alternate courses are laid lengthways then across the wall.

HOW TO LAY BRICKS AND BLOCKS

Although bricks are being laid here, the same principles apply to laying walling blocks.

1 All walls require a footing. The one shown here is a for a low wall just one brick wide: for larger and thicker walls the dimensions of the footing will have to be increased.

Excavate a trench about 30cm (12in) deep, and place about 13cm (5in) of consolidated hardcore in the bottom. Drive pegs in so that the tops are at the final height of the base. Use a spirit-level to check levels.

2 Fill with a concrete mix of 1 part cement, 2½ parts sharp sand and 3½ parts 2cm(¾in) aggregate, and level it off with the peg tops.

3 When the concrete has hardened for a few days, lay the bricks on a bed of mortar, also place a wedge of mortar at one end of each brick to be laid. For stability, always make a pier at each end, and at intervals of about 1.8–2.4m (6–8ft) if the wall is long. Here two bricks have been laid crossways for this purpose.

4 For subsequent courses, lay a ribbon of mortar on top of the previous row, then 'butter' one end of the brick to be laid.

5 Tap level, checking constantly with a spirit-level.

6 The wall must be finished off with a coping of suitable bricks or with special coping sold for the purpose.

BOUNDARY FENCES

Fences have the great merit of being more instant than hedges and less expensive than walls. That is the reason they are so often chosen by builders for new properties, and why they are frequently chosen again when the original fences come to the end of their useful life.

Closeboard and panel fences are popular, but predictable and a little boring. There are plenty of styles to choose from, however, so select a fence appropriate to your garden design yet practical for the purposes you have in mind.

If you want privacy or animal-proofing, you will have to opt for one of the solid styles, but if it is just a boundary-marker that is needed there are many attractive fences that look stylish and won't appear oppressive in a small garden.

The names of particular fence types can vary from country to country. If you do not recognize any of the names here check with the illustrations.

Closeboard
Closeboard fencing is erected on site by nailing overlapping feather-edged boards to horizontal rails already secured to stout upright posts. It is a strong, secure fence, but not particularly attractive – especially viewed from the side with the rails.

Panels
Prefabricated panels are quick and easy to erect and a popular choice for that reason. Panels are usually about 1.8m (6ft) long and range in height from about 60cm (2ft) to 1.8m (6ft), generally in 30cm (1ft) steps. The interwoven or overlapping boards are sandwiched between a frame of sawn timber. The woven style is not as peep-proof as overlapping boards.

Interlap or hit-and-miss
This combines strength and a solid appearance with better wind-filtering than a solid fence (which can create turbulent eddies that can be damaging

ABOVE: *Closeboard fencing well covered with climbing roses.*

ABOVE: *Wattle or woven fences make an attractive background for plants.*

ABOVE: *A low wooden fence is not obtrusive and can look very attractive.*

to plants). It is constructed from square-edged boards that are nailed to the horizontal rails on alternate sides. Overlapping the edges gives more privacy, while spacing them further apart can look more decorative.

Picket

Picket fences look good in country gardens, but can also be a smart choice for a small town garden. Narrow, vertical pales are nailed to horizontal timbers, spaced about 5cm (2in) apart. You can make them yourself or buy kits with some of the laborious work done for you. The simplest shape for the top of each pale is a point, but you can make them rounded or choose a more ornate finial shape. A picket fence can be left in natural wood colour, but they look particularly smart painted white. Because they are usually relatively low, and you can see plenty of garden through the well-spaced pales, they don't dominate the garden in the same way as a tall, solid fence.

Ranch-style

Ranch-style fences consist of broad horizontal rails fixed to stout upright posts. They are usually quite low, and frequently consist of just two or three rails. White-painted wood is a popular material, but wipe-down plastic equivalents are very convincing and easy to maintain. For a small garden

they provide a clear boundary without becoming a visual obstruction. Also, rain and sun shadows are not created in the way that occurs with more solid fences.

Post and chain

This is the least obtrusive of all fences. Purely a boundary marker, it will do nothing to deter animals or children, or keep balls out of the garden, but it is a good choice if you want a fence that is hardly noticeable. You can use wooden, concrete or plastic posts and metal or plastic chains. Choose a white plastic chain if you want to make a feature of the fence, black if you want the chain to recede and blend into the background.

Chain link

Chain link is not an aesthetic choice, but it is highly practical and an effective barrier for animals. It is probably best to have a contractor erect a chain link fence, as it needs to be tensioned properly. You may like the fact that you can see through it, especially if the view beyond is attractive, but you may prefer to plant

climbers beside it to provide a better screen. Choose tough evergreens such as ivy if you want year-round screening.

Bamboo

Bamboo is a natural choice if you've created an oriental-style garden, but don't be afraid to use this type of fence for any garden style if it looks right. Bamboo fences come in many shapes and sizes, and the one you adopt will depend partly on the availability and cost of the material and partly on your creativity and skill in building this kind of fence.

ABOVE: *A fence like this just needs a supply of bamboo and skill at tying knots!*

LEFT: *A white picket fence can make the boundary a feature of the garden.*

HOW TO ERECT A FENCE

Many gardeners prefer to employ a contractor to erect or replace a fence. They will certainly make lighter work of it with their professional tools for excavating post holes, and a speed that comes with expertise, but some fences are very easy to erect yourself. Two of the easiest are panel and ranch-type fences, which are illustrated in simple steps below.

HOW TO ERECT A PANEL FENCE

1 Post spikes are an easier option than excavating holes and concreting the post in position. The cost saving on using a shorter post and no concrete will go some way towards the cost of the spike.

Use a special tool to protect the spike top, then drive it in with a sledge-hammer. Check periodically with a spirit-level to ensure it is absolutely vertical.

2 Once the spike has been driven in, insert the post and check the vertical again.

3 Lay the panel in position on the ground and mark the position of the next post. Drive in the next spike, testing for the vertical again.

4 There are various ways to fix the panels to the posts, but panel brackets are easy to use.

5 Insert the panel and nail in position, through the brackets. Insert the post at the other end and nail the panel in position at that end.

6 Check the horizontal level both before and after nailing, and make any necessary adjustments before moving on to the next panel.

7 Finish off by nailing a post cap to the top of each post. This will keep water out of the end grain of the timber and extend its life.

HOW TO ERECT A RANCH-STYLE FENCE

1 Although ranch-style fences are easy to erect the posts must be well secured in the ground. For a wooden fence, use 12.5 × 10cm (5 × 4in) posts, set at about 2m (6½ft) intervals. For additional strength add intermediate posts. A size of 9cm (3½in) square is adequate for these.

Make sure the posts go at least 45cm (18in) into the ground.

2 Concrete the posts into position, then screw or nail the planks in position, making sure fixings are rust-proof. Use a spirit-level to make sure the planks are horizontal. Butt-join the planks in the centre of a post, but try to stagger the joints on each row so that there is not a weak point in the fence.

3 Fit a post cap. This improves the appearance and also protects the posts. Paint with a good quality paint recommended for outdoor use.

ABOVE LEFT: *Panel fences are easily erected and provide a peep-proof barrier, but are best clothed with plants to soften the effect.*
ABOVE: *Ranch-style fences make an unobtrusive barrier – ideal where the garden merges into the countryside.*

THE PLASTIC ALTERNATIVE

There are many plastic ranch-style fences. They will vary slightly in the way they are assembled. Detailed instructions should come with them, however, and you should have no difficulty.

The 'planks' are sometimes available in different widths – 10cm (4in) and 15cm (6in) for example – and these help to create different visual appearances. Gates made from the same material are also available from some manufacturers.

Posts are usually concreted into the ground, and the cellular plastic planks are push-fitted into slots or special fittings. Special union pieces are used to join lengths, and post caps are usually glued and pushed into position.

White ranch fencing needs to be kept clean to look good, and plastic can simply be washed when it looks grubby.

FINISHING TOUCHES

A SMALL GARDEN SHOULD BE FULL OF SURPRISES, packed with finishing touches that compensate for the lack of scope offered by limited size.

Many of the focal point techniques used in large gardens can be scaled down and applied on a small scale, and even in a small space the garden can express the owner's sense of fun and personality in the little extras that are grafted onto the basic design.

The whole area can be made to work, every corner can be exploited with devices if not plants, and a degree of flexibility can be built in that makes variety a real possibility.

ABOVE: *A seat like this suggests a gardener with a strong sense of design.*

OPPOSITE ABOVE: *Ornaments have been used to excellent effect here. A sundial commands centre stage and the eye is taken across the garden to a figure which adds light and life.*

OPPOSITE BELOW: *Figures usually look best framed by plants.*

LEFT: *This quiet corner has been transformed by white-painted trellis and seat.*

In a large garden most ornaments, furniture and fixtures like garden lights are a static part of the design. In a small garden a slight rearrangement of the furniture, the changed position of a light, or the simple exchange of one ornament for another according to mood and season means that the garden need never be predictable despite limitations of size.

Ornaments in particular can set a tone for the garden: serious or frivolous, classic or modern. They suggest the owner's taste . . . and even sense of humour. Just as the painting on the living-room wall or the ornaments on the sideboard can tell you a lot about the occupier, so garden ornaments reveal the personality of the garden maker.

Garden lighting can be practical and even a useful security measure, but it also offers scope for artistic interpretation. Experiment with spotlights in various positions and discover the dramatically different effects created by the use of light and shadows from different angles.

Arches and pergolas are a more permanent element of the garden's design, but they don't have to be planned in at the design stage and are easily added to an existing garden.

PERGOLAS AND ARCHES

A sense of height is important even in a small garden. Unless there is vertical use of plants or upright garden features, the centre of the garden will be flat. Attention will pass over the centre and go instead to the edges of the garden: exactly the lifeless effect you want to avoid.

Small trees, wall shrubs and climbers can provide the necessary verticals, but if these are in short supply an arch or pergola may be the answer.

Traditionally, and especially in cottage gardens, they have been made from rustic poles, but where they adjoin the house or link home with patio, sawn timber is a better choice. The various constructions described here are free-standing, and usually used as plant supports. Their visual effect is to take the eye to further down the garden.

If a pergola or arch seems inappropriate, similar construction techniques can be used to create an intimate arbour.

HOW TO ASSEMBLE AN ARCH

The simplest way to make an arch is to use a kit, which only needs assembling.

1 First establish the post positions, allowing 30cm (1ft) between the edge of the path and post, so that plants do not obstruct the path.

2 Fence spikes are the easiest way to fix the posts. Drive them in using a protective dolly. Check frequently with a spirit-level. Insert the posts and tighten the spikes around them. Alternatively, excavate four holes, each to the depth of 60cm (2ft).

3 Position the legs of the arch in the holes. Fill in with the excavated earth, and compact.

4 Lay the halves of the overhead beams on a flat surface, and carefully screw the joint together with rust-proof screws.

5 Fit the overhead beams to the posts – in this example they slot into the tops of the posts and are nailed in place.

HOW TO JOIN RUSTIC POLES

Rustic arches and pergolas look particularly attractive covered with roses or other climbers. You can be creative with the designs, but the same few basic joints shown here are all that you will need.

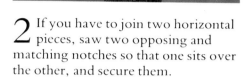

1 To fix horizontal poles to vertical ones, saw a notch of a suitable size for the horizontal piece to fit snugly.

2 If you have to join two horizontal pieces, saw two opposing and matching notches so that one sits over the other, and secure them.

3 To fix cross-pieces to horizontals or uprights, remove a V-shaped notch using a chisel if necessary to achieve a snug fit, then nail into place with rust-proof nails.

4 Use halving joints where two pieces cross. Make two saw cuts half way through the pole, then remove the waste timber with a chisel.

5 Secure the joint with a nail. For extra strength, paint the joint with woodworking adhesive first.

6 Bird's mouth joints are useful for connecting horizontal or diagonal pieces to uprights. Cut out a V-shaped notch about 3cm (1in) deep, and saw the other piece to match. Use a chisel to achieve a good fit.

7 Try out the assembly on the ground, then insert the uprights in prepared holes and make sure these are secure before adding any horizontal or top pieces. Most pieces can be nailed together, but screw any sections subject to stress.

ABOVE: *Rustic poles are an appropriate choice for a feature such as this.*

USING ORNAMENTS

Ornaments can be used around the garden in much the same way as around the house. Choose them simply because you like them, because they will look good in a particular position, or as a device for attracting attention and admiration.

In a small garden their use as a focal point is paramount. Large focal points are impractical or can only ever be few in number, but small ornaments, birdbaths, sundials, and attractive urns can be used liberally. The only 'rule' is not to have more than a couple in view at once, as they will then compete for attention rather than taking centre stage. There is no limit to the number you can use in a small garden provided they form part of a journey of discovery. Use them among plants that you only discover from a particular viewpoint, or around a corner that is not visible from where you viewed the previous focal point.

Never let ornaments detract from major focal points that form part of the basic design, and don't allow the garden to look cluttered. Aim for simplicity with surprises.

RIGHT: *A bird bath is a better choice than a sundial for a position often in shade.*

BELOW: *A chimney pot makes an unusual plinth for a sundial.*

Sundials

Whole books are written about sundials, and purists expect them to be functional. Setting them up accurately not only demands a sunny spot but quite a lot of calculations too, with compensation for geographical position. Most of us, using the sundial simply as an ornament, are happy to go out at noon on a sunny day in summer and align the gnomon to give the appropriate reading. It won't be accurate as the seasons change, but then you are unlikely to be using it to decide when it is time to leave for the office.

Accuracy may not be important, but a sunny position is. The whole object of a sundial is lost in a shady spot, where a birdbath would serve a similar design function without looking incongruous.

Choose the plinth carefully – they vary in style and height (you could even build your own from bricks) and go for a fairly low plinth if the area available is quite small.

The best place for a sundial is as a centrepiece for a formal garden, perhaps in the centre of a herb garden with paths radiating out from the centre. The lawn is another practical choice, but if the lawn is small, consider placing the sundial to one side rather than in the centre.

Birdbaths

The positions suggested for a sundial are also appropriate for a birdbath, but birdbaths are much more useful for shady positions – though not too close to trees, otherwise they become filled with leaves and debris. A birdbath can even look effective in a flower border, with much of the plinth covered by flowers, or try it as a patio ornament if you want the pleasures of watching the birds drinking from it and bathing.

Sculptures

The use of sculptures and artistic objects demands confidence. Few people react adversely to a sundial or birdbath, but sculptures or artistic ornaments that generate admiration in one person can be abhorrent to another. This should never deter anyone from using ornaments that please them, but they are bound to be somewhat more difficult to place in a small garden.

Human figures

Busts sound unlikely ornaments for a small garden, but provided they aren't too large they can look great in an alcove or on a plinth in a dull corner. Small figures can sometimes work well if surrounded by clever planting.

Animal figures

Animal figures are always a safe bet, especially if set among the plants, or even on the lawn.

Abstract ornaments

Abstract ornaments should be used with restraint – they make a considerable impact. Too many will tend to make the area look more like an art gallery than a garden.

Wall masks, plaques and gargoyles

These are great for relieving a dull wall, but are almost always best set amid the leaves of a climber such as ivy. The foliage frames the feature, and emphasizes its roles as an unusual focal point.

Gnomes

You probably love them or hate them, and that is the problem with using gnomes. One or two little people cleverly used with restraint can be very effective and add a sense of fun, but usually either they are banished from the garden or there is a whole army of them. The problem with the latter approach is that the garden will simply appear as a setting for a gnome collection.

Plinths and pedestals

Plinths are essential for raising a sundial, birdbath or bust to an appropriate height, but they can look stark in a small garden. Make more of a feature of a plinth by planting some low-growing plants around the base, and then use a few tall ground cover plants that can gradually stretch up around the base.

A plinth can look severe on a lawn and mowing around it can be difficult. Try setting one in a gravel bed with alpines around the base, or leave the bed as soil and plant thymes or other low-growing aromatic herbs.

ABOVE: *Small animal figures creeping out from the plants add a sense of fun.*

ABOVE: *This kind of ornament needs careful placing in a small garden – always take time to consider position.*

RIGHT: *Figures are often more exciting when they are discovered among the plants.*

GARDEN LIGHTING

Garden lights not only make your garden look more dramatic as dusk falls, they also extend the hours during which you can enjoy it. If you like entertaining in the garden on summer evenings, or just want to sit and relax, lights will add another dimension to the space.

When illuminating your garden you are not attempting to fill the garden with floodlights, but rather to use spotlights to pick out a particular tree, highlight an ornament, or bring to life the droplets of a cascade or fountain.

You don't even need elaborate mains lighting. Low-voltage lighting supplied from a transformer indoors is perfectly adequate for most lighting jobs in a small garden.

Lighting beds
Summer bedding looks good with pools of light thrown downwards onto the beds. If you find the lights obtrusive during the day, choose a low-voltage type that is easy to move around. Simply push the spiked supports into the bed when you want to use the garden in the evening.

Picking out plants
Use a spotlight to pick out one or two striking plants that will form focal points in the evening. The white bark of a birch tree, perhaps underplanted with white impatiens, the tall ramrod spikes of red hot pokers (kniphofias), or a spiky yucca, make excellent focal points picked out in a spotlight. Tall feathery plants, such as fennel, also illuminate well.

Spotlighting ornaments
Ornaments and containers full of plants also make striking features to pick out in a spotlight.

Before highlighting an ornament, try moving the beam around. Quite different effects can be achieved by directing it upwards or downwards, and side lighting creates a very different effect to straight-on illumination.

ABOVE: *An illuminated garden can become magical as dusk falls, and you will derive many more hours of pleasure from being able to sit out in the evening.*

Illuminating water
Underwater lighting is popular and you can buy special sealed lamps designed to be submerged or to float, but the effect can be disappointing if the water is murky or if algae grows thickly on the lenses. A simple white spotlight playing on moving water is often the most effective.

THINKING OF THE NEIGHBOURS

There is a problem with using garden lights in a small garden: you have to consider neighbours. It is unsociable to fix a spotlight where the beam not only illuminates your favourite tree but also falls on the windows of your neighbour's house. If you direct beams downwards rather than upwards, the pools of light should not obtrude.

WHEN PROFESSIONAL ADVICE IS NEEDED...

Low voltage lighting is designed for DIY installation, but 110-120 AC/DC demands special care. If you are going to wire your garden, use special outdoor fittings, and be aware of any regulations concerning the depth cables have to be buried and the protection required; you may be able to do the wiring yourself. But if in the slightest doubt use a professional electrician. If you want to keep the cost down, offer to do the labouring, such as digging trenches, yourself.

LEFT: *The best garden lighting is not obtrusive or unattractive during the day, and throws off white light when illuminated.*

HOW TO INSTALL LOW-VOLTAGE LIGHTING

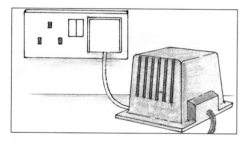

1 A low-voltage lighting kit will come with a transformer. This must always be positioned in a dry place indoors or in a garage or outbuilding.

2 Drill a hole through the window frame or wall, just wide enough to take the cable. Fill in any gaps afterwards, using a mastic or other waterproof filler.

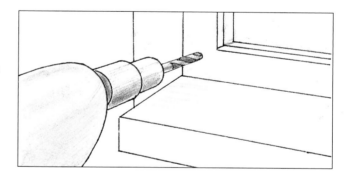

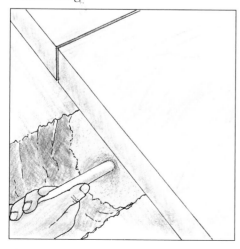

3 Although the cable carries a low voltage and you will not be electrocuted, it is still a potential hazard if left lying on the surface where someone might trip over it. Unless the lights are to be positioned close to where the cable emerges from indoors, run it underground in a conduit.

4 Most low-voltage lighting systems are designed so that the lamps are easy to position and to move around. Many of them can just be pushed into the ground wherever you choose to place them.

FURNISHING THE GARDEN

A few seats and a table make the garden an inviting place to eat, or to sit and relax. Unfortunately where space is at a premium every item has to be chosen and placed with care. Built–in seats, and especially tree seats, are a good choice for a small garden.

Portable furniture

Furniture that can be moved is useful for a quick scene change and helps to prevent your garden becoming predictable. It is surprising how effective a canvas 'director's chair' can look on a summer's day, and it is quick and easy to fold up and store when not in use.

Built–in

Built–in furniture saves space and helps prevent a small garden looking cluttered. The best place for built–in seating is the patio, where it can often be designed along with the rest of the structure. White-painted planks look smart, and can quickly be transformed with cushions to look elegant as well as feel comfortable.

Built–round

A tree seat makes an eye-catching garden feature, and this is one occasion when the advice not to have a seat beneath a tree can be ignored! White paint will help the seat to stand out in the shade of its branches.

Wrought and cast iron

Genuine cast and wrought iron furniture is expensive and very heavy, but alloy imitations are available with all the charm of the original but at a more manageable price and weight. White is again a popular colour, but bear in mind that although this type of furniture can stand outside through-out the year, it will soon become dirty. Cleaning the intricate patterns isn't easy. Colours such as green look smart yet don't show the dirt.

Use cushions to add patches of colour, and to make the chairs less uncomfortable to sit on!

LEFT: *White-painted metal furniture looks tasteful and can help enliven a dull corner of the garden.*

BELOW: *A charming wooden seat.*

BELOW LEFT: *A reconstituted stone seat has a timeless appeal that beckons you to sit and rest.*

Wooden seats and benches

Timber seats can be left in natural wood colour to blend with the background or painted so that they become a focal point. White is popular, but green and even red can look very smart. Yacht paint is weather-resistant.

Plastic

Don't dismiss plastic. Certainly there are plenty of cheap and nasty pieces of garden furniture made from this material, but the better pieces can look very stylish for a patio in the setting of a modern garden.

HOW TO MAKE A TREE SEAT

1 Start by securing the legs in position. Use 3.8cm × 7.5cm (1½in × 3in) softwood, treated with a preservative. You will need eight lengths about 68cm (27in) long. Concrete them into position.

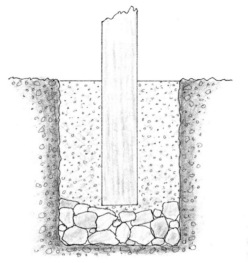

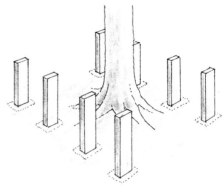

2 Position the legs about 38–45cm (15–18in) apart in two rows about the same distance either side of the trunk.

3 Cut four pieces of 2.5 x 5cm (1 x 2in) softwood for the cross-bars. Allow 7.5cm (3in) over-hang at each end. Drill and screw these to the posts.

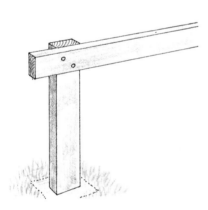

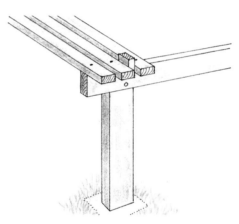

4 Then, cut slats to the required length (the number will depend on the size of your seat). Allow for a 2.5cm (1in) space between each slat. Paint the slats and cross-bars with white paint (or a wood preservative or stain if you prefer), and allow to dry before final assembly. Test the spacing, using an offcut of wood as a guide, and when satisfied that they are evenly spaced on the cross-bars, mark the positions with a pencil. Then glue and nail into position.

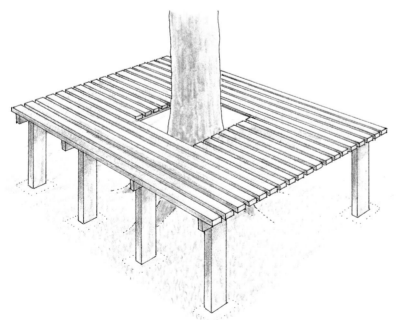

POTS FOR DOORWAY DECORATION

Always choose an imposing plant in an attractive container to go by the front door, and if possible one that looks good for a long period.

This is the place for a clipped bay in an ornate pot or Versailles tub, or an attractive bamboo in an oriental-style container.

If you have chosen imposing plants to go by the front door supplement these with a group of smaller containers that add seasonal colour, and perhaps scent. Don't be afraid to move pots around to maintain interest. Keep a small lilac in a tub or grow pots of hyacinths and move them to the front door as they come into flower to add a heady perfume.

Formal shrubs

If space really is limited and the rest of the garden has a formal style, a couple of clipped or trained evergreens can look elegant throughout the year. Clipped bays are good, but in cold areas are likely to suffer from damaged leaves in winter, but many conifers have a naturally formal outline and remain attractive throughout the year with minimal attention. Box can be bought clipped into topiary shapes, and though expensive to buy will add instant impact. You can easily buy a box plant and clip it into a ball or pyramid

shape over the course of a couple of years, if you are happy with a simple geometric shape.

Scented delights

Scent always arouses comment from visitors to the door. In winter you will have to rely on bulbs like hyacinths and *Iris danfordiae*. In spring follow these with daphne and then lilac (both indifferent for the rest of the year, so be prepared to move them to a less conspicuous part of the garden after flowering).

Summer brings the opportunity to use scented bedding plants such as flowering tobacco plants and stocks.

Climbers in pots

A climber round the door always looks attractive, and you can usually erect a trellis for support. If there is a choice, plant directly into the ground, but if that is not possible, pot a climber in a tub. Large-flowered

LEFT: *Remember to appeal to the sense of smell as well as sight. Here lavender not only colour co-ordinates, it adds a touch of fragrance as well.*

BELOW LEFT: *Formally clipped box can be expensive to buy, but with patience you can train your own. They are ideal for a formal setting.*

BELOW: *Don't forget that pots can always be used to grow well-trained shrubs.*

clematis will do well, and even a honeysuckle. You can try a climbing or rambling rose, but these are more demanding in pots.

GROUPING POTS AND PLANTS

If isolated pots seem to lack impact, try grouping them together – the mutual support they lend each other gives them a strength that they lack individually. If the pots are rather plain, placing smaller ones in front will mask those behind and bring the display almost to ground level.

Groups in the porch

Make a bold display in a porch by using tall plants, especially evergreen shrubs, at the back and smaller flowering plants in front.

If space is limited, instead of going for a lush effect with lots of foliage and flowers, concentrate on the containers rather than the plants. Decorative pots are often available as matching sets. Grouping these together looks good even if the plants they contain are only mediocre.

Groups in corners

Difficult corners are an ideal place in which to use containers to create colour, filling in a spare piece of ground where nothing much seems to do well. Patios usually have corners that would otherwise remain unused. Group shrubs or tall houseplants at the back and colourful summer bedding plants in front, along with bright-leaved indoor plants for the warmest months.

Alternatively, choose a small group of elegant containers and use the plants in a more restrained way. A trailer growing from a pedestal container with a cluster of distinctive small pots around the base can be as eye-catching as a large group.

In a dull corner, perhaps formed where two wooden fences join, or where house joins fence in a sunless position, try making a bed of small-sized gravel on which to place a group of terracotta pots. Red gravel will help to bring colour. Fill the pots with bright annuals for the summer, and winter-flowering pansies and bulbs for winter and spring. Try spacing the pots out and adding a few interesting pieces of rock among them.

ABOVE LEFT: *Grouping plants in a porch makes a high-impact feature. Replace plants when they have passed their best, to keep it looking good.*

ABOVE: *Feature groups of plants in containers where the garden needs an uplift. The beach pebbles add an individual touch.*

LEFT: *Individually, these containers would not look special, but grouping them makes a focal point.*

Groups on the lawn

Clusters of pots are an ideal means to breaking up a large expanse of lawn. Don't stand them directly on the grass, but use a bed of sand or gravel – this will stand out well from the grass, and make mowing round the containers easier.

YEAR-ROUND CONTAINERS

In a large garden containers are usually used for splashes of summer colour. The voids left in winter when the plants have died are not so noticeable among the many other garden features. In a small garden, and especially on a patio, bare containers in winter look positively off-putting, and only emphasize the lack of year-round plants.

The choice for summer is limitless, so the emphasis here is on autumn and winter – the seasons for which most effort has to be made.

Year-round troughs and boxes

Dwarf evergreen shrubs and dwarf conifers, in their many shapes and colours, will provide year-round interest. But to prevent them becoming so much background greenery, leave space in front to plant a few bulbs or small bedding plants. Allow for a space the size of a small pot for these seasonal plantings, so that you can easily replace the small flowers as they finish. Grow a reserve of them in pots to fit the space.

Autumn highlights

Grow one or two autumn-glory shrubs in tubs that you can bring out of their place of hiding when you need a final burst of colour.

Ceratostigma willmottianum has compact growth and lovely autumn foliage tints while still producing blue flowers. Berries can also be used as a feature, and you can usually buy compact pernettyas already bearing berries in your garden centre.

Winter colour

Some winter-flowering shrubs can be used in tubs, such as *Viburnum tinus* and *Mahonia* 'Charity'. But try being bold with short-term pot plants like Cape heathers (*Erica* × *hiemalis* and *E. gracilis*) and winter cherries (*Solanum capsicastrum*) and similar species and hybrids. You will have to throw them away afterwards, but they will look respectable for a few weeks even in cold and frosty winter weather.

LEFT: Solanum capsicastrum *is widely sold as a houseplant in the winter months, but you can use it as a short-term plant to add a touch of colour to permanent plantings of evergreens in outdoor containers. Those pictured were still happy in late winter. Discard once the berries shrivel.*
BELOW: *The intense blue flowers of ceratostigma last well into autumn, when there is the bonus of rich foliage colour before the leaves fall.*

HOW TO MAKE A ROCK GARDEN

1 The base is a good place to dispose of rubble, which you can then cover with garden soil – the ideal place for soil excavated from the pond.

2 It is best to use a special soil mixture for the top 15–23cm (6–9in), especially if soil excavated from the pond is used. Mix together equal parts soil, coarse grit and peat (or peat substitute), and spread this evenly over the mound.

3 Lay the first rocks at the base, trying to keep the strata running in the same direction.

4 Lever the next row of rocks into position. Use rollers and levers to move them.

5 As each layer is built up, add more of the soil mixture, and consolidate it around the rocks.

6 Ensure that the sides all slope inwards, and make the top reasonably flat rather than building it into a pinnacle. Position the plants, then cover the exposed soil with a thin layer of horticultural grit.

THE PLANTING

*Hard landscaping (paving, walls, fences, pergolas, and so on)
is what gives a garden a strong sense of design, and provides the skeleton
that gives the garden its shape. But it is the soft landscaping – the plants –
that provides the flesh, shape and texture of the garden.
The same basic design can look very different in the hands of
gardeners with different ideas on the use of plants.*

ABOVE: *Mixing different types of plant can
be very effective. This border contains shrubs,
herbaceous plants, bulbs, and grasses.*

OPPOSITE: *No matter how attractive the
design of a garden, it is the plants
that make it pretty.*

BEDS AND BORDERS

BEDS AND BORDERS NEED TO BE PLANNED. THE shape will affect the overall appearance, of course, but there are also practical considerations such as the amount of maintenance required, the theme to be created, as well as the crucial question of the actual plants to be used.

Formal beds and borders are normally dictated by the basic design concept, which will often determine the type of plants you can use. A formal rose garden will clearly feature roses, and only the 'filler' plants might have to be debated. A classic style with neat symmetrical beds cut into the lawn, or edged by clipped box, demands the type of formal bedding associated with this type of garden.

Herbaceous and shrub borders are much more open to interpretation, and the actual plants used will have as much affect on the overall impression created as the shape or size of the border.

In traditional large gardens there is a clear distinction between herbaceous borders and shrub borders, but few small gardens can afford this luxury and the inclusion of a 'mixed border' is the usual compromise. Here shrubs jostle for

LEFT: *By curving the corners of borders in a small garden you can generate extra planting space that helps to make the garden more interesting.*

OPPOSITE ABOVE: *A garden like this, with plenty of shrubs such as roses, require little maintenance and because the hard landscaping is minimal is relatively inexpensive to create.*

OPPOSITE BELOW: *Single-sided herbaceous borders can look right in a rural setting if you have enough space. A border like this can be colourful for many months.*

position with herbaceous plants and annuals, while summer bedding plants and spring bulbs make bids for any areas of inhabitable space left. There is nothing wrong with this type of gardening: the border looks clothed long after the herbaceous plants have died down, and there will be flowers and pockets of changing interest for a much longer period than could be achieved with shrubs alone.

Colour themes are also difficult to achieve in a small garden, and although single-colour borders can be planted in a small garden, it is best to be a little more flexible. Settle for a 'golden corner' rather than a golden border, or a blue-and-silver theme for just part of a border rather than a more extensive area.

Small beds cut into the lawn do not have to be filled with summer bedding and then replaced by spring bulbs and spring bedding. Instead plant them with blocks of perennial ground cover, or use a perennial edging and plant seasonal flowers within it.

ISLAND BEDS

Traditionally, low-growing seasonal plants have been grown in beds cut into the lawn – island beds – and taller herbaceous plants and shrubs placed in long borders designed to be viewed from one side. Island beds planted with herbaceous plants and shrubs bridge this divide, and provide planting opportunities that can be put to good use in a small garden.

Planting principles

Island beds are intended to be viewed from all sides, so the tallest plants usually go in the centre and the smaller ones around the edge. Don't be too rigid, however. Concentrate on creating a bed that you have to walk around to see the other side, rather than simply planting tall summer flowers like delphiniums in the centre. Shrubby plants, even medium-sized evergreens, might be better for the centre of the bed, with other lower-growing shrubs creating bays that can be filled with plants that die down for the winter. Your bed will then retain its function of breaking up a lawn and creating a diversion that has to be explored.

Don't be afraid to plant a small tree, such as *Malus floribunda*, in an island bed, to create much-needed height.

If seasonal bedding appeals more than shrubs and border perennials, then island beds can still be used creatively for these.

The question of shape

Most people think of island beds as informal in outline, but you can introduce rectangular beds if this suits the style of your garden.

Curved beds generally look much more pleasing, however, especially if you introduce broad and narrow areas so that there are gentle bays.

Design considerations

Use an island bed to break the line of sight. By taking it across the garden, an island bed may distract attention from an uninspiring view – whether beyond the garden or simply the fence itself. Attention is directed to the sides, and as you walk around the bed, the eye is taken into the bed rather than to the perimeters.

A series of island beds can be used to divide up a long, narrow garden. Instead of the eye being taken in a straight line to the end, the beds become a series of diversions.

BELOW: *Island beds help to break up a large lawn, and create a sense of height.*

ONE-SIDED BORDERS

Single-sided borders are useful if you want to create flowery boundaries around the perimeter and emphasize an open space within the garden, turning the garden in on itself. These borders are also useful for taking the eye to a distant focal point, and, by varying the width of the border, you can create a false sense of perspective that can appear to alter the size of the garden.

Straight and narrow beds
Most gardens have at least some straight and narrow borders around the edge of the lawn, a favourite spot for roses or seasonal bedding. If you want to cut down on the regular replanting work, plant with dwarf shrubs as backbone plants then include flowering ground cover herbaceous plants such as hardy geraniums and spring bulbs to provide flowers over a long period.

Make a border look wider by laying a mowing edge. Then use plants that will sprawl over the edge, softening the hard line and giving the impression of a wider border.

The advantages of curved borders
Straight edges are easier to mow and trim, but unless the border is wide and variation is created with the use of shrubs of various sizes, they can appear unimaginative and may take the eye too quickly along the garden, making it seem smaller. Gentle curves that create bays enable the plants to be brought further out into the garden and provide much more adventurous planting scope.

It may be possible to modify an existing straight border by cutting into the lawn. Bear in mind that mowing time is likely to be increased rather than decreased, however.

ABOVE: *Single-sided borders are the best choice for small town gardens that have high enclosing walls, especially if you can use climbers or tall plants to hide the wall.*
LEFT: *A single-sided mixed border.*

Turning corners
Don't forget that borders can turn corners. Right-angled turns seldom look satisfactory, however, so add a curve to the corner. This will give it greater depth at that point.

You can even take the border right round the garden in a continuous strip. A small square garden with a circular central lawn surrounded by border can look quite spectacular if well planted with a wide range of plants that hold interest throughout the seasons.

HOW TO MAKE BEDS AND BORDERS

If you are making a garden from scratch, areas allocated to lawns, beds and borders will be laid out accordingly, but you can often improve an existing garden by altering the shape of a border, or creating beds in what is currently a large and uninspiring lawn.

HOW TO MARK OUT AN OVAL BED

For small formal beds, such as ovals and circles, it is best to sow or lay the grass over the whole area first, then cut out the beds once the grass has become established.

Start by marking out a rectangle that will contain the oval. Afterwards you can check that it is square by measuring across the diagonals, which should be the same length.

Place a peg half way along each side, and stretch a string between them. The two strings will cross at the centre point. Then cut a piece of string half the *length* of the oval and using a side peg as a pivot, insert pegs where it intersects the long string along the centre.

Make a loop from a piece of string twice the distance between one of these two pegs and the top or bottom of the oval (whichever is the furthest away).

With the loop draped over the inner pegs, scribe a line in the grass while keeping the string taut. You can make the line more visible by using a narrow-necked bottle filled with dry sand instead of a stick.

Use an edging iron to cut out the shape, then lift the grass with a spade.

HOW TO MAKE A CURVED BORDER

1 If you want a quick and easy method, and can trust your eye for an even curve, lay hosepipe where you think the new edge should be. Run warm water through it first if the weather is cold, otherwise it may not be flexible enough to lie on the ground without awkward kinks.

2 The best way to judge whether the curves are satisfactory is to view the garden from an upstairs window, and have someone on the ground who can make further adjustments if necessary.

3 When the profile is satisfactory, run sand along the marker (dry sand in a wine bottle is a convenient method). Use an edging iron to cut the new edge, then lift the surplus grass and dig the soil thoroughly before attempting to replant.

4 An alternative and a more accurate way to achieve smooth curves is to use a stick or bottle fixed to a string attached to a peg. Use this as a pivot. By adjusting the length of the string and the position of the pivot, a series of curves can be achieved. Cut the edge as before.

HOW TO GET A NEAT EDGE

Emphasize the profile of your beds and borders, as well as your paths, by giving them a crisp or interesting edge. A mowing edge is a practical solution for a straight-edged border. Curved beds and borders usually have to be edged in other ways.

Some methods, like the corrugated edging strip and the wooden edge shown below are not particularly elegant, but they help to prevent the gradual erosion of the lawn through constant trimming and cutting back, and they maintain a crisp profile.

Using ornate or unusual edgings
For a period garden, choose a suitable edging. Victorian-style rope edging tiles are appropriate. If you live in a coastal area, consider using large seashells. If you enjoy your wine as well as your garden, why not put the empty bottles to use by forming an edging with them? Bury them neck-down in a single or double row, with just a portion showing.

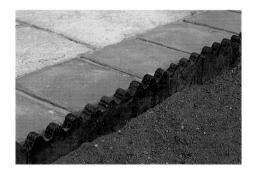

TOP: *It is possible to buy a modern version of Victorian rope-edging.*
ABOVE: *Edgings such as this are useful if you want to create a formal or old-fashioned effect.*

HOW TO FIT EDGING STRIPS

Edging strips like this are available in a thin metal, soft enough to cut with old scissors, or in plastic. These strips help stop erosion of the grass through frequent edge clipping and cutting back. Although these may not be the most decorative edging strips, they are quick and easy to fit.

1 Make a slit trench along the lawn edge with a spade, then lay the strip alongside the trench and cut to length. Place the edging strip loosely into it.

2 Backfill with soil for a firm fit. Press the strip in gently as you proceed. Finish off by tapping it level with a hammer over a straight-edged piece of wood.

HOW TO FIT WOODEN EDGING ROLL

Wired rolls of sawn logs can make a strong and attractive edging where you want the bed to be raised slightly above the lawn, but bear in mind that it may be difficult to mow right up to the edge.

1 Cut the roll to length using wire-cutters or strong pliers to cut through the wires, and insert the edging in a shallow trench. Join pieces by wiring them together. Backfill with soil for a firm fit. Make sure that the edging is level, first by eye. Use a hammer over a straight-edged piece of wood to tap it down. Then check the height with a spirit-level. Adjust as necessary.

HOW TO PLANT A BORDER

You don't have to be an artist to draw a functional planting plan. You can buy simple computer programs that will help you draw one up, but you still have to provide the plant knowledge that makes a border come alive and fulfil your own expectations. You can achieve results that are just as acceptable, and probably just as quickly, with pencil and paper.

A SCALE OUTLINE

1 Draw the outline shape of the bed and border, marking on the scale. Use graph paper so that you can easily estimate the size of a particular plant as you work.

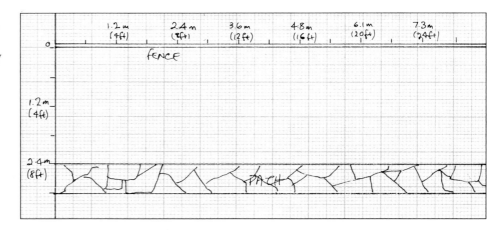

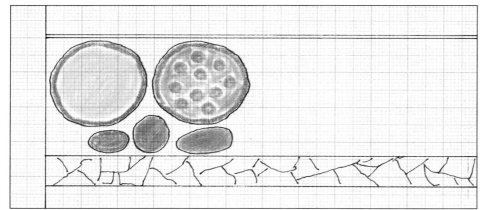

2 Make a list of plants that you want to include. Be sure to add essential details such as height, spread, and flowering season. If you find it easier to move around pieces of paper rather than use pencil and eraser initially, cut out several pieces of paper of appropriate size, with the height and flowering period marked on. You could colour them – evergreen greens, variegated green and gold stripes, and flowering plants in the colour of the blooms.

3 Either start with a basic plan with a series of spaces to be allocated (just indicate whether tall, medium or small), or shuffle around your cut-outs until they appear to form a pleasing pattern. Don't worry about whether the plants will fill the exact shape – with time they will all grow into each other, and in the meantime you can fill the gaps with annuals.

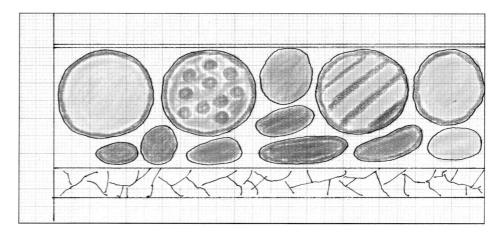

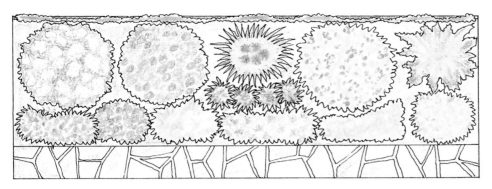

4 When satisfied with your key plants, draw these in on a more detailed planting plan. Then fill in the gaps with other plants, not necessarily on your priority list.

If you feel sufficiently artistically inclined, you can try a profile view that will give a better idea of how the border will look – though you can only make a snapshot of how it would look in one season.

HOW TO PLANT HERBACEOUS PLANTS

GUIDELINES TO GETTING IT RIGHT

● Unless the plants are large, plant in groups of about three – a bold spash usually looks better than single plants. Using single plants just because space is limited is a common mistake – the impact is often better if you use fewer kinds but more of each.
● Take into account the likely ultimate height, but remember that plants may grow taller in one garden than another.
● As a rule place the taller plants at the back (in the centre of an island bed), with the smallest at the front. But don't follow this too slavishly unless planting formal summer bedding. A few focal point plants that stand out from the rest can be very effective.
● Consider planting the border so that different parts are at their best at different times, perhaps starting with spring flowers at one end and working through to autumn at the other.
● Use foliage plants to maintain interest throughout the border.

1 Always prepare the soil first. Dig it deeply, remove weeds, and incorporate a fertilizer and garden compost if impoverished. Most herbaceous plants are sold in pots, so space them out according to your plan. Change positions if associations don't look right.

2 Water thoroughly about half an hour before knocking the plant from its pot, then remove a planting hole with a trowel. If the roots are wound tightly around the root-ball, carefully tease out a few of them first. Work methodically from the back of the border, or from one end.

3 Firm the soil around the roots to remove any large pockets of air.

4 Always water thoroughly after planting, and keep well watered in dry weather for the first few months.

PLANTING FOR TEXTURE

Quite dramatic plantings can be achieved simply by planting blocks of the same plant – whether summer bedding, herbaceous perennials or shrubs. If the garden is seen as an area of voids and masses, blocks of colours and textures, the overall impression can be as important as individual plants. Ground cover plants are ideal for this purpose.

ABOVE: *Thyme is a useful ground cover for a sunny position, and will even tolerate being walked upon occasionally.*

CONVENTIONAL PLANTING

Many ground cover plants spread by sideways growth, sending up new plants a short distance from the parent. These are best planted like normal herbaceous or shrubby plants. Suppress weeds initially with a 5cm (2in) layer of a mulch such as chipped bark. This is also the best way to plant any kind of ground cover that forms part of a mixed border.

HOW TO PLANT GROUND COVER

If planting ground cover plants as a 'texture block', or perhaps to cover an area of ground that is difficult to cultivate, such as a steep slope, it is best to plant through a mulching sheet. You can use black polythene, but a proper mulching sheet is better as it allows water to penetrate. However, do not use the sheet method for plants that colonize by spreading shoots that send up new plants, as the sheet will prevent growth by suppressing the shoots as effectively as the weeds.

1 Prepare the ground well, eliminating weeds. Add rotted manure or garden compost, and rake in fertilizer if the soil is impoverished.
Secure the sheet around each edge. Tuck the edges firmly into the ground and cover with soil. Make two slits in the form of a cross where you want to plant.

2 Plant through the sheet as you would normally, firming the soil around the roots.
If you use small plants, planting with a trowel will not be a problem. Water thoroughly.

3 Although the mulching sheet will suppress weeds very effectively while the ground cover is still young and not able to do the job itself, it does not look attractive, so cover it with an ornamental mulch, such as chipped bark.

HOW TO PLANT SHRUBS

1 Most shrubs are sold in pots, and can be planted at any time of the year when the ground is not frozen or waterlogged. Space them out in their pots first, then adjust if the spacing does not look even.

2 Prepare the ground thoroughly, making sure it is free of weeds. Dig in plenty of organic material such as well-rotted manure or garden compost. Otherwise use a proprietary planting mix.

3 Excavate the hole and try the plant for size. Use a garden cane or piece of wood across the hole to make sure the plant is at its original depth. Add or remove soil as necessary.

4 Remove the plant from the pot. If the roots are tightly wound around the root-ball, carefully tease some of them free, to encourage rapid rooting.

5 Firm the plant in well to eliminate large air pockets. Gentle pressure with the heel is an efficient way to do this, or alternatively you can do this by hand.

6 Rake or hoe in a balanced fertilizer to get the plant off to a good start. In autumn use one that is slow acting or has controlled release, to avoid stimulating growth during the cold months. If planting in winter, wait until spring before adding the fertilizer. Water well, then mulch with a 5cm (2in) layer of organic material such as garden compost, cocoa shells, or chipped bark.

INDEX